EXTREME FOOD

EXTREME FOOD

What to eat when your life depends on it

BEAR GRYLLS

wm

WILLIAM MORROW
An Imprint of HarperCollins*Publishers*

First published in Great Britain in 2014 by Bantam Press, an imprint of Transworld Publishers.

EXTREME FOOD. Copyright © 2014 by Bear Grylls Ventures. All rights reserved. Printed in the United States of America. No part of this book may be used or reproduced in any manner whatsoever without written permission except in the case of brief quotations embodied in critical articles and reviews. For information address HarperCollins Publishers, 195 Broadway, New York, NY 10007.

HarperCollins books may be purchased for educational, business, or sales promotional use. For information please e-mail the Special Markets Department at SPsales@harpercollins.com.

FIRST WILLIAM MORROW PAPERBACK EDITION PUBLISHED 2015.

Designed by Julia Lloyd

Library of Congress Cataloging-in-Publication Data has been applied for.

ISBN 978-0-06-241675-9

15 16 17 18 19   OV/RRD   10 9 8 7 6 5 4 3 2 1

To the one and only Danny Cane.

Ex-Parachute Regiment legend and the man
who first taught me to eat worms!

Who would have guessed . . .

CONTENTS

INTRODUCTION

I've eaten some pretty rough things in my time. Eyeballs, testicles, raw animal flesh, warm blood, the stomach contents of a freshly killed animal. Some have made me gag, some have surprised me and some, if truth be told, I'd go to the ends of the earth never to have to taste ever again.

But I don't regret a single mouthful.

Because when it comes to the hard realities of survival, you have to think about food in a different way. It's no longer just about taste. It's now about energy, fuel and, ultimately, life.

Over the course of my travels, I've met plenty of people who have found themselves in genuinely terrifying survival situations. Men and women who would have died if they hadn't summoned up reserves of courage and ingenuity they never knew they had.

All of them have one thing in common: they never thought it would happen to them. They never thought that *they* would be the ones who would have to survive when nature decided to stack all the cards against them.

And when that moment came, one of their most difficult challenges was finding the food they needed to give them energy and keep up their morale.

It's hardly surprising. Most people live a life very far removed from the natural world. And that's especially true when it comes to the food we eat. It often arrives on our plate pre-packaged and ready made. We rarely pick our own plants, catch our own fish or kill our own meat. We think of insects, amphibians and reptiles as either frightening or gross, which means we ignore half the food with which the world can supply us in favour of meals that are safe, comfortable and easy.

No wonder, then, that the pursuit of food can be one of the greatest challenges we face when the odds are against us in the fight for survival.

This book is not about the safe, the comfortable or the easy option. I want it to teach you how to fend for yourself when it comes to dinner time in the wild, no matter where you are in the world.

By the end, you'll have learned how to do the simplest things, like cooking up a feast with your buddies round the campfire with raw materials that you might have brought with you from home, through to the extreme end of wild food survival – I'm talking blood, gore, beasts and bugs (anyone for fried tarantula?). You'll have learned how to hunt and kill wild animals for food, and how to prepare them so that you can eat them safely. You'll have learned how to catch and eat insects and amphibians, crocs and snakes. How to get your teeth into food you might never have thought of as being food in the first place.

Some of the food in this book might turn your stomach. I make no apology for that. I always say that survival is rarely pretty, and that's especially true when it comes to survival food. But at the end of the day, is there really that much difference between eating a slug and eating a chicken? Between eating a slice of frog meat and a rasher of bacon? It's just a question of what you're used to.

Ultimately, I want to show you that there are few things more empowering than knowing how to use what nature has to offer. Because one day, that knowledge could save you and your loved ones' lives.

Bear.

EXTREME FOOD

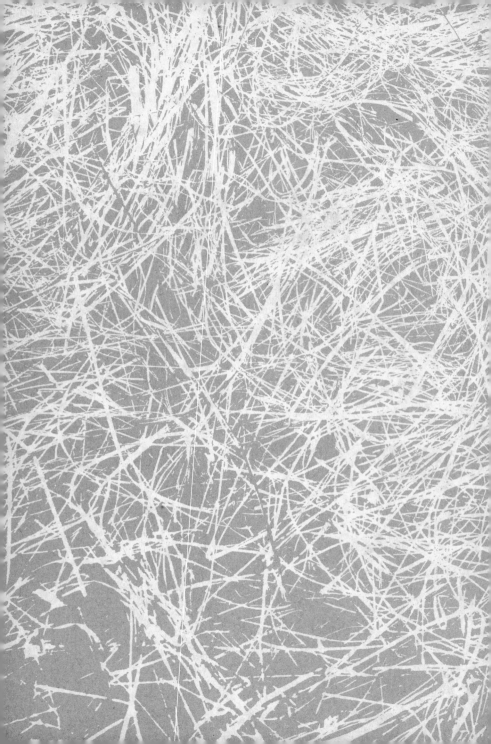

PART ONE
THE BASICS

1

COOKING UP A STORM ROUND THE CAMPFIRE

Later in this book you're going to learn how to live off the land, and how to use what nature has given us to survive when we're miles away from the comforts of a kitchen.

But eating in the wild isn't always a matter of fighting for survival. Sometimes we put ourselves out on the trail for the sheer fun of it. On these occasions we can find ourselves better prepared when it comes to trail food. So, to kick off, we're going to look at some great ways of getting fuel inside you when you're hiking or camping with your buddies.

Some of these ideas revolve around ingredients that you can easily pack and bring with you. I'm not going to ask you to forage, catch or kill anything just yet! But that's OK, because knowing how to prepare yourself before an expedition is just as important as knowing how to survive in the wild during the aftermath of an expedition gone wrong.

And remember this: cooking round the campfire is always fun, but it can also teach you some important lessons. Learning how to build a fire, or making sure your water is safe to drink, or understanding the differences between cooking out in the open and in a well-equipped kitchen – these are all important techniques for anyone who wants to stand a chance of surviving for real in the wild.

Learn them well now and you'll be in a much better position when things get a little edgier . . .

NUTRITION IN A NUTSHELL

Food is fuel. The harder we push our body and the more we expect of it, the more fuel we need. If you're going to function properly in the wild, you'd better make sure the tank doesn't get empty. And you tend to get out what you put in.

Except it's not quite as simple as that. Take bananas, for instance – they're a great, high-energy food. But if you lived off nothing but bananas, you'd go downhill pretty fast. The body needs a varied diet to continue functioning. That means making sure you consume all the major food groups: fats, carbohydrates, proteins, vitamins and minerals. So before we start out, let's establish what these food groups are and where you find them.

FATS

Fats in general get a bad press, but in fact they are essential to survival because certain vitamins need fats in order to be absorbed. Without fat in your diet, you'll die. You *should* try to make sure, however, that your diet is heavier on the 'good' fats and lighter on the 'bad' fats. Good fats (mono- and polyunsaturated fats) are found in fish, nuts and vegetable oils. Bad fats (saturated fats), which can raise your cholesterol and lead to heart disease, are found in meat and dairy products.

When you're exerting yourself in the wild, it's worth remembering that you get twice as much energy from fats as you do from carbohydrates or proteins. And if you're going to burn that energy, it's OK to put it in the tank.

CARBOHYDRATES

These are divided into 'simple' carbohydrates (basically sugars) and 'complex' carbohydrates (whole grains, pulses, nuts, root vegetables). The complex carbohydrates are the ones you want. Sugars burn fast and quickly, whereas complex carbohydrates release their energy slowly over a long period of time. It's the difference between burning a scrunched-up piece of newspaper and a nice chunky log. Which one's going to keep you warmer for longer?

PROTEINS

You get the same amount of energy from protein as you do from carbohydrates, and your body needs it to keep your cells and muscles healthy. At home, you're best off trying to get plenty of protein from whole, natural foods such as avocados, pulses and nuts, but in the wild your main sources of protein will be from any animals you manage to kill or insects you manage to catch. The humble earthworm, for example, is extremely high in protein. But more on that later!

VITAMINS AND MINERALS

These are found in tiny quantities in the body, but are essential for it to function properly and most of them come from the food you eat. Your body stores about a month's worth of vitamins, but after that these reserves will start to deteriorate. Vitamin C is the first to go: the result is scurvy, the scourge of sailors on long voyages before people understood about vitamins – it can result in bleeding gums, loose teeth, jaundice and eventually death. If you eat a varied diet of plant matter and a little good-quality, grass-fed meat, most of your vitamins and minerals will be replenished, and it's worth remembering that they can come from the strangest sources. The Inuit derive much of their Vitamin C from caribou liver and seal brain.

Depending on your size and sex, you burn in the region of 2,000 calories a day if you're just sitting around doing nothing. If you're moving about and exerting yourself in the wild, you need a load more. Your body will use up its own reserves if you put nothing in the tank. After that, how much energy you have depends wholly on how much you eat. 1g of fat has 9 calories. 1g of either carbohydrate or protein has 4 calories.

WATER

As I'm sure you'll have figured by now, water is essential to survival. Far more important than food. So if you're setting out on the trail, the first thing you should pack is your water bottles. And when you *have* water, you should ration it carefully. Drink just enough to keep yourself hydrated, but not so much that you just excrete the excess as sweat or pee. Treat your water as your most precious resource, because that's exactly what it is.

The trouble with water is that it's very heavy. There's a limit to the amount you can carry, and if you're going to use it for cooking, chances are you'll need to find another water source to supplement it.

Your best, and most abundant, sources of clear water are rivers and lakes. But sometimes such rivers and lakes aren't available. In that case you have to be resourceful. Even a muddy puddle can supply you with water. Digging down on the outside of dry river beds, or under the sand around greenery in deserts, can lead you to damp soil.

But before you drink water from any source in the wild, you must make absolutely sure that you clarify and purify it first.

A sponge and a plastic bag are great items to keep in your pack. If you find yourself far from water, you can use the sponge to soak up any rain that falls, then squeeze it into the plastic bag. This was the method used by US Air Force pilot Scott O'Grady when he was shot down behind enemy lines in Bosnia in 1995. He also survived for almost a week by eating grass, leaves and insects – more on those later!

Transporting water in the field can be a challenge if you don't have a water bottle with you. For that reason, consider carrying a couple of condoms in your pack. You'd be surprised how strong they are and how much water they can hold when fully stretched and placed inside a sock for support (just don't overfill them – a burst condom's no good to anyone!). They're also great for stashing away bits of tinder and keeping it dry (see page 25).

CLARIFYING WATER

In the wild, the water you come across will not always be crystal clear. If it's cloudy, or even downright muddy, you'll need to filter out all the nasty stuff before you purify it.

You can buy special water filters for use in the field. These are simply bottles with a disposable filter inside, and they're available in different capacities. If you know you're going to have to clarify water, these are great bits of kit to take along with you.

Or you can make your own. All you need is a cloth bag (or even the bottom of a trouser leg, tied at one end) and a piece of string. You'll need to fill the bag with filtering material that you gather on the trail. Think fine sand, coarse sand, tiny stones and larger pebbles, or even grass. Layer this material in the bag, with the fine material at the bottom, gradually moving up to the coarser material at the top.

Use the string to tie the bag to the limb of a tree. Pour the cloudy water into the top and collect the clear water that drips from the bottom.

Remember: *you still need to purify the water once you've filtered it.*

During the jungle training phase of UK Special Forces selection, a component of the belt kit is a Millbank bag – a fabric bag for filtering water in the field. If you have one of these, you don't need to fill it with filtering material, but here's a word of advice: make sure it's thoroughly soaked before you add the water to be filtered, and let the first few centimetres of water filter through before you start collecting it. And, of course, always purify the water once it's filtered.

PURIFYING WATER

The water that you've taken from a fast-moving stream might *look* fantastically clear and pure, but it can harbour some very nasty things. I've suffered badly in the past from drinking contaminated water, so you can take it from me: dysentery, stomach cramps and bloody poo are not what you want in the wild.

There are two ways to purify clear water. One is to take with you some water-purifying tablets which are made of chlorine or iodine. They might make the water taste a bit odd, but you can add a neutralizing tablet to counteract this – although there's no doubt that a funny taste is a lot better than the alternative.

Or you can boil it. At sea level, water needs to be boiled for a minute to make it safe to drink. For every 300 metres you find yourself above sea level, boil for an extra minute.

SOME GREAT WAYS OF BOILING WATER IN THE FIELD

Being able to boil water is a crucial part of survival cooking. Not only does it enable you to purify your drinking water, but the process of boiling certain foods renders them edible by destroying bacteria and parasites. But sometimes doing this in the field isn't straightforward. On page 105 you'll find a clever way of using coals and hot rocks, but here are a few other ideas to help you get that all-important boiling water.

A storm kettle

Storm kettles are awesome. The best-known are called Kelly Kettles. The kettle itself is a hollow metal cylinder that acts as a chimney. The skin of the cylinder contains your water. You light a little fire in the base – using small twigs or whatever else you can find – which burns really well because of the chimney and heats the water in the cylinder in just a few minutes. Storm kettles come in all sorts of sizes, but the smaller ones are quite light and easy to carry when you're on the trail.

A plastic bottle

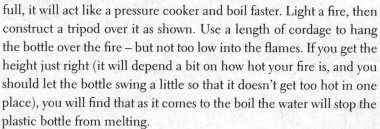

It seems like anywhere you go, chances are there's an empty plastic bottle littering the ground or shoreline. But if it's clean, it can be put to good use in a surprising way.

To do this, first fill your plastic bottle to the top with clear water, then put the cap back on. If your bottle is really full, it will act like a pressure cooker and boil faster. Light a fire, then construct a tripod over it as shown. Use a length of cordage to hang the bottle over the fire – but not too low into the flames. If you get the height just right (it will depend a bit on how hot your fire is, and you should let the bottle swing a little so that it doesn't get too hot in one place), you will find that as it comes to the boil the water will stop the plastic bottle from melting.

This probably isn't something you'd want to do every day, because the plastic can potentially release certain chemicals into the water. But if you've no other way of boiling water, it's a great technique to have up your sleeve (not to mention a good party trick around the campfire!).

A birch bark pot

As you'll see elsewhere in this book, in a survival situation the birch tree is one of your best friends. (You'll probably recognize a birch tree when you see one by its white, papery bark with horizontal black markings.) On page 62 you'll find some ways of using it to make a nutritious survival drink, but you can also use the bark to make a container for boiling water.

Get yourself a roughly rectangular piece of birch bark. It needs to be at least the size of a piece of A4 paper so that the pot's not too fiddly to make and holds a decent amount of water when it's finished. Fold it as shown. (The diagonal folds have to crease inwards.) Make some pegs by splitting the ends of some small sticks and use these to hold the folds in place. The resulting pot should hold water without any leaking.

To boil the water, you can place the full birch bark pot directly on the embers of your fire (no flames!). As with the plastic bottle trick, the water should stop the birch from burning as it comes to the boil.

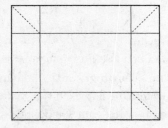 

FIRE

It may be that you have your own trail stove with you. If so, great. But a real fire is a lot more fun, and knowing how to light one is a great skill to have.

Throughout this book, you're going to come across some clever ways of using fire to cook your food. There are lots of different methods of building outdoor ovens, or of boiling and grilling your food. But they all rely on being able to get a decent fire going in the first place. So here are the basics.

A fire needs three things in order to burn: fuel, oxygen and heat. Take away any one of these elements and you've got no fire. Always make sure you've gathered plenty of fuel before you light your fire. Don't pack the fuel so tightly that the oxygen fails to circulate. And always start with small pieces of fuel first, so that you build up enough heat to get the larger pieces burning. If you rush things, your fire *will* go out. End of story.

FUEL, AND HOW TO LIGHT IT

You'll need to think of your fuel in three groups: tinder, kindling and larger fuel. Tinder gets the fire started and kindling builds up the heat so that you can add the larger fuel to keep the fire going without putting it out.

Tinder

You can use all sorts of material for tinder. It just needs to be small, fine and easy to ignite with a match. Dry grass, pine cones, wood shavings and old birds' nests all make good tinder.

You'll need something with which to light it, of course. A match is fine – just make sure you keep your matches dry. Otherwise, a cigarette lighter is a great idea.

Alternatively, you can take a 9v battery and a wad of very fine wire wool. As long as it's fine enough, if you touch the wool to the two terminals of the battery it will heat up very quickly – enough to get your tinder lit.

I have often also used the foil of a chewing-gum wrapper and an AA battery to make fire. Cut a long strip of foil about

5mm in width, then cut out a V-shape halfway along the strip. Connect one end of the foil to each end of the battery. The foil will heat up very quickly, ignite the paper backing – and you have fire!

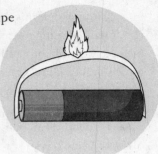

For a clever way of using a mushroom as tinder, see page 81.

Kindling

Your best source of kindling is dead branches, but try to find some that are still on the tree rather than those that have fallen to the ground. They'll be drier. (A good test of whether kindling or wood is suitable to burn is to snap it. If it sounds like a fire crackling, you know it's good to go.) You want your kindling to be pencil-thin and about 40cm long. Slowly feed the dry kindling on to the lit tinder, taking care not to smother it.

Larger fuel

If you have an axe, then cut some larger branches – again, as dry as possible – into different-sized pieces so that you can gradually increase the size of the fuel you add to your fire. If you don't have an axe, try to gather wood of different sizes. If you want a fast, merry blaze, keep your fuel smaller; if you're after more of a slow-burner, add larger pieces of fuel, but only when the fire is hot enough to take them.

A GOOD FIRE FOR COOKING

First, prepare your site. Make sure you're not too close to any bushes, dry grass or overhanging trees.

If the ground is wet, muddy or snow-covered, build a base of thick green lengths of wood to raise the fire off the damp ground.

Find some rocks to make a U-shaped perimeter and place a large rock at one end, facing into the wind. This will make the fire easier to light and will also lessen the risk of it spreading.

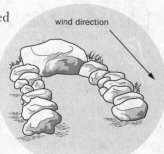

wind direction

Now cover the fire area with tinder before placing your kindling over the top in a criss-cross pattern. For a cooking fire, this is better than a teepee-style arrangement because it helps you get the fire going over the whole area of the fire pit.

Light your tinder. When the kindling is burning nicely, start adding your firewood. Try to use pieces of about the same size, and spread them out all over the fire pit. When the flames burn down and you're left with white coals, get yourself a decent stick and arrange the coals so that you have a big pile at the back, a medium pile in the middle and a small pile at the front. This gives you three different zones, each with a different amount of heat, so you can control how fast or slow your food cooks.

Lots of the ideas that follow suggest cooking food in the embers of the fire, but if you have a grill you can place it across the perimeter rocks now. Try to make sure that it's stable and flat, so that your pots and pans don't slip around.

TWO GOLDEN RULES

1. Always choose your fire site carefully. Are there any overhanging branches that might ignite if a spark hits them? Are you in an exposed position where the wind might cause the fire to spread? Have you prepared the ground well, brushing away any flammable material? Remember that the only stuff you want to burn is what *you* put on the fire.

2. When you've finished with your fire, have you carefully and properly extinguished it? Use water if you can, and plenty of it – or pee on it! – and make sure that there are no underground roots still smouldering. Otherwise you can smother it using sand or soil. Either way, you mustn't leave the site until you are *absolutely sure* that your fire is out. Remember: it only takes a tiny spark to burn down an entire forest . . . (Or, as I once told Stephen Fry, it only takes one little ember to start a fire. To which he replied, as he watched me peeing on the fire, 'And one little member to put it out!')

KEEPING FOOD COOL

Later in this book you'll learn some ways of preserving food that you've caught or hunted in the wild (see Chapter 10). But as you've probably worked out from the big white thing in your kitchen, the best way to keep food fresh is to keep it cold. That can be a bit of a challenge out in the field, so here are some ideas to keep perishable items cool.

NATURE'S REFRIGERATOR

Running water is always cold. If you have some watertight boxes, you can submerge them in a fast-running stream to keep the contents chilled. To make sure the current doesn't wash them away, either place a sufficiently heavy rock on top of the box, or create a U-shape of rocks pointing in the direction of the current so that it traps the items you want to chill. Alternatively, tie boxes or bottles to a fixed point, like a tree on the bank of the stream, before submerging them in the running water.

HAY BOXES

Again, keep your food in a sealed box. Dig a hole in the ground bigger than the volume of the box. Place the box in the hole, then stuff the gaps around it with hay or dried grass and cover it with hay. This will insulate the box and its contents and keep them cool.

KEEP COOL UNDER CANVAS

It can get pretty hot under canvas when the sun's out. You need to be aware of this and move your food around to the cooler parts of your tent as the day wears on. Before you go to bed, put your food on the west side of the tent so that it doesn't get heated up by the rising sun.

SOME BASIC CAMPFIRE FOODS

You can buy all sorts of foil-packed, dehydrated or freeze-dried trail foods for when you're out in the field with your friends. And don't get me wrong: in a survival situation they can make the difference between life and death, as anyone in the military who has had to depend on an MRE (meal, ready-to-eat) will tell you.

But several days of dehydrated food can get very boring and sap your morale. It's impossible to overestimate how much a hot meal, freshly cooked over your campfire, can boost both body and soul. And for most people, the whole ritual of preparing food out in the open is the most awesome part of a camping trip or an expedition into the wild.

So here are a few ideas for some good, basic recipes and techniques that you should build into your repertoire. This kind of food is never going to be haute cuisine. It's going to be ballsy, hearty food, a bit rough round the edges – and all the better for that!

TRAIL SPICES

If you want to stop your trail food from being bland, you should think about making up your own trail spice box. It doesn't need to take up a lot of space or be very heavy, because when it comes to spices, a little goes a long way. Empty Tic Tac boxes are good for storing your individual spices, which you should keep in a waterproof plastic box. You should definitely consider including salt, ground black pepper, curry powder and cinnamon. Paprika, mustard or chilli powder will give your food a kick. After that, it's up to you.

You should also think about taking a small bottle of olive oil and a small bottle of honey. Make the bottles plastic – they're lighter to carry and won't shatter.

BANNOCK

Bannock is a kind of bread that you can easily cook over an open fire. It has been eaten by explorers, soldiers and indigenous peoples for centuries. It's delicious, nutritious and easy to make in the field. Plus it's fun to prepare.

There are probably as many recipes for bannock as there are wild-food enthusiasts. Some people cook it in frying pans, but you can just as easily cook it on sticks. I'm going to tell you how to do both. The great thing about bannock is that you can prepare the basic mix at home, then just add a little water out in the field when you're ready to cook it.

Basic bannock mix

125g wholewheat flour (or white or a mixture of white and wholewheat)
1 tsp baking powder
a good pinch of salt
30g dry milk powder
1 tbsp lard

Mix together the dry ingredients, then rub in the lard until you have a mixture that looks like breadcrumbs. Pour this into one of those zip-lock freezer bags (maybe use two – one inside the other – if you're going on a long trek) and pack it away carefully with the rest of your stuff.

THE FRYING PAN METHOD

Put your pan over the fire or stove (be really careful, because if the frying pan has a metal handle it can get really hot). Even better, you can nestle the frying pan into the embers of your campfire if they're not too hot. When you're ready to cook your bannock, open up the zip-lock bag and trickle a little water into it. Squish it all together and keep adding water little by little until you get a good, thick dough. Pull it out of the bag and shape it into a flat round the right size for your frying pan, then lay it in the pan. Now let the bannock cook through (it will probably rise a bit). How long it takes depends on how hot your fire is, but you're probably talking about 15 minutes. Shake the pan now and then to stop the bannock from sticking. And when it's cooked, tuck in!

THE STICK METHOD

This is a lot of fun. First you need to find a good stick – not too thick and heavy, because you're going to be holding it over a fire for about 10 minutes. Three fingers thick is about right, and it needs to be long enough that you don't have to sit too near the fire.

Peel off any outer layers of bark from the stick, then hold the end near the fire so that it 'tempers' – gets good and hot without burning or charring. This means that the bannock will cook from the inside as well as from the outside.

Prepare your bannock mix by adding water to the zip-lock bag as above. Remove from the bag and fashion the dough into a long sausage shape.

Now, wrap your dough-sausage round the stick, squishing it as you go – you don't want the dough to be more than 1cm thick or it won't cook through properly.

To cook your bannock, find a point above the fire where you can comfortably hold your hand for about 10 seconds. Any closer and the outside of the dough will cook before the inside. Hold the dough-wrapped stick over the fire and rotate it like a spit for about 10 minutes. You'll know it's cooked when the bannock can be slipped easily off the stick.

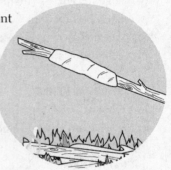

You can add all sorts of things to the basic bannock mix: spices, herbs, raisins, even chocolate drops. Failing any of these, once it's been cooked you could sprinkle it with a little cinnamon from your trail spice box. Experiment, enjoy and trust me – after a long day in the field, everyone will love it.

CAMPFIRE PORRIDGE

When you're out on the trail, you need to be careful about what you eat. A chocolate bar shoved into your pack might seem like a good idea when you're hungry, but the energy you get from the sugar will burn away very quickly. You need food whose energy will last a little longer. Porridge is that food. It'll give you slow-release energy that lasts a long time.

Like bannock mix, porridge oats are a brilliant foodstuff to take with you out on the trail. Keep them dry and they last for ages, which means that as long as you have a supply of water and a campfire, you'll have a quick and easy way of making yourself a hot, filling and very nutritious meal that will give you plenty of energy.

Although porridge is normally made from oats, you can in fact make it from all sorts of grains: quinoa, cracked wheat and bulgur wheat are all good. To make your porridge, you'll need two parts water to one part grain. It's a good idea to measure out your dry ingredients before you leave home (use a cup or vessel that you'll have with you in the field so that you then know exactly how much water to add). Put the grain into a zip-lock bag and mark it with the amount of water you need.

To cook the porridge, bring your water up to the boil in a small trail saucepan, then add a little salt. Slowly stir the oats in (use a spoon if you want, but a clean stick will be fine) and carry on stirring until the porridge starts to thicken and bubble. Make sure you keep scraping the bottom of the pan to stop it sticking.

You can eat the porridge like that, or you can add some flavourings. Raisins and nuts are good (keep them in different zip-lock bags to the oats if you're bringing them from home). Edible wild berries are even better.

And remember, oats make flapjacks, which are great 'on the go' snacks and will give you much more sustained energy than just a chocolate bar. Honey, oats and butter/olive oil = tasty, slow-release power. If you're making them on the trail, don't get too hung up about a recipe. Just mix the quantities that seem right. And don't worry if it doesn't turn into proper bars – the mixture tastes just as good eaten warm with a spoon.

PANCAKES

You can buy packets of ready-made pancake mix, which are less healthy, but there's no reason not to make your own trail pancakes from scratch. Obviously it's not that easy to take perishable items like eggs and milk with you, but it's not hard to take a supply of powdered egg and powdered milk. Pack some wholewheat flour (get those zip-lock bags out again!) and you have everything you need to make some fantastic campfire pancakes. If you can turn these out for everyone at breakfast, you'll be the most popular person you know!

You need about a cupful of flour and a couple of teaspoons each of powdered egg and powdered milk – but don't stress about the proportions too much. Stir in enough water to make a batter about the same consistency as single cream. Get your pan good and hot, add a little drop of oil and enough of the batter to cover the bottom of the pan, cook for a minute or so and then flip the pancake and cook for another 30 seconds. Don't worry if it tears slightly – it all adds to the charm. You can eat it like that, or add a bit of sugar or honey. A sprinkling of cinnamon from your trail spice box, if you have one, will take it to a whole new level.

Pancakes are also great for wrapping round other foods to make a more substantial meal. Think grilled meat, fish, vegetables or fruit: anything you might enclose in a bread wrap you can also put in a pancake!

BREAKFAST IN A PAPER BAG

Carrying a frying pan around with you on the trail can be heavy. But a paper bag weighs almost nothing, and you can use it to cook your breakfast. Layer rashers of bacon inside the bag, making sure that the bottom of the bag is completely covered and that the bacon extends up the sides a little. Now crack an egg or two into this little bacon 'boat'. Roll down the top of the bag a few turns and place it on a grate over the campfire (but keep it away from any actual flames). The fat from the bacon stops the bag burning, and when you see that it's absorbed about halfway up the bag, breakfast's ready. Cool, eh?

CAMPFIRE POTATOES

Potatoes are very filling, so if other foods are in scarce supply you can still fill up. To cook them on the campfire, tin foil is your best friend.

You don't need to take a big bulky roll; just tear off a sheet at home, maybe half a metre in length, and fold it down into a neat square to take with you. When you've used it, you can wipe it clean and recycle it for another meal.

The easiest way to cook a potato is to wipe it clean, wrap it in a couple of layers of tin foil and place it in the embers of your fire. It should cook through in about 45 minutes, depending on the size of the potato and the heat of the fire. But it doesn't take much to make your potato a bit more special, if you're on a short trip and able to bring with you an onion and some grated cheese. Try slicing the potato thinly and placing it on a sheet of tinfoil. Slice the onion up thinly too, add it to the potato and sprinkle the cheese over. Add a little water so that everything can steam nicely while it cooks. Season with whatever you fancy from your trail spice box (you'll definitely want some salt and pepper), then gather up the tin foil and scrunch it tightly so that the steam doesn't escape. Put the package in the embers of the fire and let it cook for about 45 minutes.

DUTCH-OVEN STEWS

A Dutch oven is a heavy casserole, usually made of cast iron, with a tight-fitting flat lid. Sometimes they have little legs, and often a handle that you can use to suspend the oven over a fire (like you see in cowboy movies). Dutch ovens are great because the thick cast iron, although it absorbs heat slowly, retains it well and spreads it evenly, so there's less chance of burning your food.

The most common way of using a Dutch oven is to place it in the embers of your fire, then lay hot coals and ash on top of it, which allows the food to cook from the top as well as from the bottom and the sides. You can also get a tripod from which you hang the oven, enabling you to control the heat by altering the height above the fire.

Because of its weight, you're unlikely to end up taking a Dutch oven on a backpacking trip, but if you're setting up a fixed camp, it's a bit like taking your kitchen with you. You can use your oven to make almost anything, including breads and desserts, but it really comes into its own when you want to make a hearty campfire stew. Later on in this book, in Part Two, you're going to learn how to hunt and butcher your own wild meat. I can't think of many game animals that wouldn't benefit from an hour or two simmering in a Dutch oven on the campfire. If you get used to cooking with one of these ovens, you'll be amazed at how you can turn a few unpromising-looking ingredients into an outdoor meal to remember.

Here are a few pointers on how to make a basic Dutch-oven stew. Once you've got the idea, you can adapt it to suit whatever food you manage to lay your hands on. Because remember: surviving outdoors is all about improvising with what you have to the best of your abilities.

Meat stew

To make this very simple stew, you'll need some cubes of meat, some carrots, onions and potatoes cut up into chunks, some water and some salt and pepper. You want your chunks to be about 2cm square, but don't worry too much about the size. I can't tell you how much of each – it depends on what you have with you and how many people you have to feed. Just go with it.

Place the Dutch oven in the embers of your fire and let it get up to temperature. You want it so that you can (carefully!) place your hand just inside the oven for about 5 seconds before it gets too hot. Now add your meat cubes and enough water to just about cover them. (If you've got a stock cube or two knocking around, you could add those too.) Put the lid back on the oven then place some hot embers or coals on the top. Cook until the cubes of meat are tender, which will probably take about an hour. Brush the embers from the top of the oven, remove the lid and add your vegetables and plenty of salt and pepper. Replace the lid and the embers, then cook for another 20 minutes until the vegetables are soft. Now serve up your stew and eat it while it's good and hot.

Once you've mastered the basic idea, you can add all sorts of things to your campfire casserole: chunks of bacon, tins of tomatoes or beans, mushrooms, pretty much any vegetable. Read the chapter on wild plants (see pages 38–39) and you'll find lots of delicious wild greens that you could stir in at the last minute. Try out whatever inspires you – it will probably be delicious!

One other tip: you can use the flat lid of a Dutch oven like a griddle – brilliant for cooking your breakfast pancakes.

Campfire Popcorn

Everyone loves popcorn, right? Popcorn kernels are dead easy to carry around with you and won't spoil if you're on the trail for a long time. What's more, popcorn is incredibly healthy and a lot of fun to make on a campfire. Here's how.

Get yourself a nice big sheet of tin foil – about 40cm by 40cm. Fold your foil in half, then scrunch up the two ends so you have an open 'boat' of foil. Now add a good handful of popcorn kernels and a splash of oil (about 2 tbsp). Seal the edges of the tin foil so you have a loose, enclosed pouch with enough room for the corn to pop and expand. Thread a piece of string through the top of the pouch (soak the string in water first if you're worried it might burn) and tie it so that the pouch can dangle from the string, then tie the other end to a long stick. You can now dangle the pouch of popcorn over your campfire. You'll soon hear it start to pop, and it's done when the popping stops. Open up the pouch and sprinkle the fresh popcorn with a little salt or sugar, whichever you prefer. Or try dribbling honey, or melted chocolate, on it for a more indulgent version, and eat with a spoon!

2

WILD PLANTS

Wherever plants can grow, there's the opportunity for food. If you know what you're looking for, the natural world is an inexhaustible larder of nutrition. Not that you need me to tell you that – you've been blackberry-picking before, right? Even the most unexpected plants can be eaten if you're really struggling to find food. Ordinary grass, for example, doesn't taste too bad and it's full of nutrients. Sometimes in a survival situation, you just have to go with what you've got.

That's not to say, of course, that you can eat *anything* that grows. You can't. Mix a few berries from the deadly nightshade plant (the clue's in the name) into your pancakes and it will probably be the last meal you ever eat. So, knowing how to forage for your own food isn't just a matter of identifying what's edible, it's also a matter of being able to recognize those plants that can put you out of action for days, or even kill you.

Learning how to live off the land is an essential skill for anyone who wants to be prepared in case they ever find themselves in a survival situation. But it's also a technique that will make your trips into the wild much more fulfilling. Once you know where to look, you'll start seeing free food everywhere. You'll probably even find yourself foraging for goodies to bring home with you.

It would be impossible for me to list all the edible plants out there. Instead, I'm going to suggest a few that are common in temperate, tropical, desert and Arctic regions. Once you've got used to finding, cooking and eating some of these, I hope you'll be encouraged to start learning more about the wild larder and the amazing variety of foods the natural world can offer us, even in the most inhospitable parts of the world.

And remember, the best time to practise your survival skills is *before* you find yourself in a survival situation, not when your back's against the wall. Foraging for edible plants is a good example of a skill you can learn wherever you are and whenever you want, and which could very well save your skin at some point in the future.

With that in mind, I'm also going to point out a few highly poisonous plants that you absolutely must *never* eat. Again, it's not exhaustive by any means, but I hope it will make you understand that there are some plant baddies out there. They might look pretty, or even appetizing, but you really don't want them in your gut.

But what if you're in a survival situation and you don't know if a plant is edible or not? Later in this chapter (page 57), you'll find the edibility test that every good soldier knows – a way of testing whether a food is toxic when your life depends on it – as well as some instructions on what to do if you think you or your buddies have poisoned themselves.

BEAR'S GOLDEN RULES FOR COLLECTING WILD PLANTS

1. If you're new to foraging, choose plants that are easily recognizable. Some can be readily confused with poisonous species. Don't run before you can walk.

2. Cross-reference your identification sources. ***Don't just rely on this book***. Consult several field guides from your local library, or use the internet. Better still, go out into the field with someone who,

literally, knows their onions. Make sure you've positively identified a plant before you even think about eating it. Remember that the knowledge of how to survive in the field doesn't come overnight. You accumulate it over a lifetime. If you learn to identify one of these plants every few weeks, you're doing pretty well.

3. Wash your plants in clean water before you eat them, and don't eat polluted plants. That means: avoid plants that are growing close to busy roads, where wild animals might have brushed past them, or where dogs or other creatures may have done their business. (Animal urine really isn't a good condiment!)

4. Don't be greedy. Harvest only what you need and no more. If we're going to live off the land, we have to make sure that what we do is sustainable.

5. A good rule of thumb is that the younger a plant is when you cook it, the more palatable it will be. Don't cook your harvest for *too* long unless otherwise stated. The more you cook it, the fewer nutrients you'll have at the end. Remember: this isn't just food, it's fuel for your body as well.

TEMPERATE REGIONS

The following plants all grow abundantly – but not exclusively – in temperate regions.

STINGING NETTLES

It might sound a bit weird eating the leaves of a plant that every kid is taught to avoid from the moment they learn to walk. But don't worry!

Once you cook nettles, the sting vanishes.

It's a good idea to wear a pair of rubber gloves and a long-sleeved shirt when you're picking them, though. If you stumble across them when you don't have any gloves, a good tip is to put your hand inside a plastic bag. The best bits are the young leaves (once they start to flower, you've missed the boat), so you'll need to strip them from the stalks by running your hand from the top down.

Nettles are one of the most delicious wild foods out there. They're certainly the king of wild greens. People have been eating them for hundreds of years, and not only do they taste good, they're also one of the richest sources of iron in the natural world. They say that eating nettles makes your hair brighter and shinier, and herbalists use them to clear up eczema and other skin conditions.

If you're using them in the kitchen, you need to wash the leaves well and steam them over boiling water. You can use them in any recipe that calls for spinach, and if you pick and cook a load, you can freeze any leftovers for later.

A really great, warming way to use nettles in the wild is to cook up a batch of nettle soup on the camp stove. Exactly how you make it depends on what other ingredients you have with you. If you can get your hands on an onion and a few potatoes (or, even better, go for sweet potatoes, which are my favourite and are even healthier!), then chop them up with a good sharp knife and fry them in a little olive oil. Add some water and let the soup simmer till everything is soft. Then throw in a few handfuls of washed nettle leaves. They'll quickly wilt down and the stings will disappear. Season the soup with plenty of salt and pepper from your trail spice box. If you've got a bit of nutmeg in there, so much the better. Divide the soup between your buddies and eat while it's steaming hot. That'll keep you going for a good few hours in a cold tent!

DANDELIONS

Yep, dandelions.

You'll find them everywhere in the spring and summer. Even in the dead of winter, though, you sometimes come across dandelions in sheltered places, so they can be a welcome source of food when you're really up against it.

The yellow flowers, which turn to white, feathery moons as the plant ages, are instantly recognizable. And the name dandelion comes from the French *dents de lion* – lion's teeth – because of the shape of its leaves with their jagged edge. It's the leaves you're after, especially the young ones, which can be eaten raw. They have a slightly bitter taste, and as they get bigger and older you'll probably want to boil them before eating. Dandelion leaves are said to have a diuretic effect, hence the old English name for them: pissabed!

WILD GARLIC

You'll probably smell the distinctive, pungent aroma of wild garlic before you actually see the plant – most likely in damp, shady woodland. Wild garlic bulbs put up their long, lush, pointed leaves in early spring, followed by beautiful white flowers on long, thin stems. The leaves and flowers can be eaten raw, or chopped up and added to other wild foods to give them a flavour a little milder than ordinary cultivated garlic. Wild garlic is also full of nutrients, including allicin, which has antibacterial and anti-viral properties, so it's a great plant for keeping you healthy in the field.

Wild garlic also makes a good mosquito repellent. Rub the leaves on to your skin so that you're smearing yourself with the plant's juices. (Be warned, though: this might also repel your friends!)

ACORNS

Acorns are the nut of the oak tree. Nowadays we pretty much think of them as the diet of squirrels, even though humans have used them for food for long periods throughout history. The Ancient Greeks ate them, as did the Native Americans. During the Second World War, when Allied blockades halted the Germans' supply of coffee, the Germans used ground acorns as a coffee substitute. Even nowadays acorns are sometimes ground down into flour, and they are still used in Korean cuisine.

So there's no reason why you shouldn't regard them as a survival food (they're a particularly good source of calories), or forage for them when you're out in the field. Don't just go popping an acorn into your mouth, though. They need preparation first.

Acorns are very high in tannin. This is a compound that occurs naturally in lots of foods, but acorns have a particularly high concentration, which makes them very bitter to eat. To remove the tannin, soak your acorns in warm, purified water for a few hours. You may find that some of the acorns sink and some of them float. Keep the ones that sink and discard the ones that float because this indicates that they've dried out and aren't good to eat (although it may also indicate the presence of an edible grub – see opposite). The water will turn brown as the tannin leaches out.

Now, remove an acorn from its little cup, cut off a piece of the nut and put it in your mouth. If it's still bitter and unpalatable, soak it again in fresh water, and repeat until the tannin taste has gone. Remember to remove all your acorns from their cups before eating them.

Once you have separated out the floating acorns, examine them carefully. They may have a little hole in them which indicates the presence of a tiny grub, which will wriggle out of its own accord. Don't waste it! That grub is edible.

The tannin water left over from soaking your acorns is good for treating burns and rashes in the field if you don't have access to more advanced medical treatment. Boil up a clean dressing in the tannin water, let it cool, then apply it to burns or rashes.

And once you know how to identify the above plants, move on to these: primroses, wild chicory, hawthorn, curled dock, alexanders, fat hen.

TROPICAL REGIONS

Tropical regions are incredibly fertile, especially in the areas around water sources. That means lots of plants, and lots of potential food. But you really need to know what you're eating. The most abundant plants in the tropics are palms: many of them have edible parts, so it's worth your while learning as much about them as you can.

Here are three awesome tropical survival plants.

COCONUT PALMS

The coconut palm can be a lifesaver. Not only is the 'meat' of the coconut an excellent food, but the milk inside is a good source of sterile liquid (so sterile, in fact, that it was used as a glucose substitute in IV solutions for soldiers in the Second World War) and has a high sugar and vitamin content.

You find most liquid in the green-orange unripe specimens. They grow in clusters high up the coconut tree. Getting them down can be a challenge, but there's a knack to climbing a coconut tree. Choose a tree with a good slant, position yourself on the upper side of the slope, grab the trunk with your hands and start pushing yourself up with your feet. Keep your feet flat against the tree so you get as much grip as possible, and pull against the trunk with your hands as you climb. You'll soon get the hang of it. When you reach them, the green coconuts should twist off fairly easily.

Let the coconuts fall to the ground before you climb back down. To do this, you can either reverse the process, or grab around the trunk and slide down slowly, gripping with both feet and inner thighs. But, guys, beware of getting crunched nuts!

Back on the ground, you need to hack off the husk at the bottom of the coconut with your knife so that you get a triangular point. Slice off the tip of this point to expose the creamy white flesh. It tastes amazing, and once you've cut into it you have a ready-made bowl of fresh liquid with more electrolytes than ordinary water and way more potassium than a banana. Get it down you.

It's worth knowing that coconuts have other uses in the wild. You can render oil from coconut meat by gently heating it over your fire. This oil can be used to treat sunburn and other sores. It can also be used to make an improvised torch (just pour it into a small container, add a piece of string for a wick and leave it to set), and the smoke of burning coconut oil will keep mosquitoes away.

If you find mature coconuts, with their characteristic hairy husks, you can use the husk as tinder, and even bind it together to create cordage. (And eat the coconuts, of course!)

All in all, they're a godsend.

BANANAS AND PLANTAINS

These two related plants grow wild in tropical regions. You don't need me to tell you how to eat a banana. You can eat plantains in just the same way, though the fruit can be harder so they're often boiled, baked or fried. They're a brilliant food source because they provide loads of slow-release energy and are packed full of vitamins and minerals.

Even more awesomely, if you come across a banana or plantain tree, then you've also come across a water source. Grab yourself a length of bamboo (it's often growing nearby). Sharpen one end with your knife, then jab it into the trunk of a banana tree. It will act as a tap: liquid from the inside of the tree will soon start to trickle down the bamboo. You can make yourself an improvised container by digging a hole underneath the end of the bamboo stick and pressing a banana leaf (they're massive) into it to form a bowl.

BAMBOO SHOOTS

Bamboo is just about the most versatile plant you can come across in the jungle. You can use it to build shelters and rafts, you can sharpen it to make a hunting spear. Shavings from dried bamboo stalks also make brilliant tinder. You can even cook with it (see below). Best of all, you can eat it.

The most edible parts of a bamboo plant are the young shoots at the base. They have a tough outer section, which you'll need to split with your knife to reveal the tender part within. You can eat them raw (they might be quite bitter), but there may be some risk of toxicity so if you can, it would be better to boil or fry them.

If you come across very large bamboo stalks, which can grow up to 20cm in diameter, you can use them to make cooking vessels. You'll see that the stems have evenly spaced notches along them. These form solid cross-sections. If you make a cut just outside two adjacent notches, you'll have a tube closed at both ends. Lay the tube on its side and cut a long hole into the top. Now you have a cooking pot that you can rest on a frame (made from smaller bamboo) over your open fire. Perfect for boiling water and cooking whatever you want inside it.

Alternatively, make both your cuts just below two adjacent notches. That will give you a tube with one closed end and one open end. Lay the closed end on the ground, and prop up the open end over your fire with a forked stick. The result is a deep, tubular cooking pot. Genius.

And once you know how to identify those, move on to these: yukkas, Brazil nuts, snake gourds, breadfruit, papaya.

DESERT REGIONS

In the desert, the hunt for water is far more pressing than the hunt for food. In fact, you should never eat if you don't have enough water because it will just make you more dehydrated. The good news is that there are some desert plants that provide you with both nutrients and hydration. But don't be tempted to eat just *anything* that grows: there are cacti out there that are full of moisture but extremely toxic to humans.

PRICKLY PEAR CACTUS

You've got to be careful with cacti. Many of them produce edible fruits, while the flesh contains chemicals that will cause diarrhoea. Not so the prickly pear cactus: the fruit and the pads are both edible.

Pads first. You'll need to remove the spines, then peel off the green outer skin. The interior is edible raw. It's bland, and a bit slimy, but a great source of liquid – the pad itself is more than 80 per cent water.

Later in summer, the fruits start to mature at the tip of the pads. They're ready to eat when they've turned red, and they have a great citrus taste. Carefully cut the fruit from the plant with a knife, then split it lengthwise down the middle. You'll see a little interior flesh and plenty of seeds. They're all edible. (You can make the unripe pears more palatable and digestible by boiling them first.)

Although prickly pears are edible, there are other species of cactus that look similar and aren't. Those that aren't have a milky sap, so if you see that: avoid! (In fact, a milky sap is generally a good indicator of an inedible plant.)

ACACIA TREES

These are really widespread in desert regions, especially in Africa and Australia. There are more than a thousand different species. The young leaves and shoots can be boiled and eaten. Same goes for the seeds, which you can also roast then grind down into flour to make a kind of desert porridge.

The roots of the acacia tree are also edible, and a great source of moisture in desert climates.

WILD GOURDS

These are part of the watermelon family and grow in a sprawling mat on the ground of even the hottest deserts and wastelands. The fruits, when they're ripe, form yellow spheres about the size of a large lemon. The flesh can be very bitter, especially when under-ripe, but if you boil the fruit whole you can make them more palatable. The seeds can be boiled or roasted – they're a good source of healthy oils. The gourd plant produces a yellow flower which is edible raw, and you can chew on the stem tips to extract water.

And once you know how to identify those, move on to these: agave, date palms, desert amaranth, carob, baobabs.

ARCTIC REGIONS

Food can appear to be scarce in extremely cold regions just as in extremely hot ones. But if you know what you're looking for, you can still forage for edible plants even in these inhospitable parts of the world.

ARCTIC WILLOW

This is one of the best sources of Vitamin C in the tundra. It's a low shrub, no more than 60cm high. You can eat the young shoots, the leaves, the bark and even the roots (peel them first).

ICELAND MOSS

Most lichens are edible, including this stuff, which grows well in the otherwise forbidding Arctic regions. It is only a few centimetres high, and its colour varies from white, through grey, to a kind of reddish brown. It's very dry to the touch, but becomes more malleable when soaked – which you should do, before boiling it well. Iceland moss is a particularly good survival food because it stores really well when dried.

FIREWEED

This is the national flower of Greenland. It can grow up to 2.5 metres high and has amazing purple flowers. For the Inuit, it's a valuable food source – they eat it as a salad alongside seal and walrus blubber. Every part of it is edible. The young shoots, stems and leaves can be eaten raw or simmered in stews. The roots are also edible raw. As the plant gets older, you'll need to split the stems in half to get to the soft edible sections.

Later in the year, the seed pods become covered in a woolly fluff which you can use as tinder for starting a fire. I guess that's where the name comes from.

And once you know how to identify those, move on to these: cloudberries, reindeer moss, red and black spruce, salmonberries, bearberries.

AN AWESOME WAY OF COOKING WILD PLANTS (AND ALMOST ANYTHING ELSE!) WITHOUT POTS OR PANS

In a survival situation, you might not always have cooking equipment with you. Here's a great way of using nature to deal with that. We're going to use wood and fire to make ourselves a cooking pot.

First off, light a fire (see pages 24–8) and let it burn down so that you have plenty of glowing embers. Now find yourself a chunk of wood big enough to be turned into a decent-sized container – imagine that, if you were to scoop out a hole, it would hold a litre or two of liquid. Make sure it's not from a variety of tree that's poisonous to humans (pine, cedar, oak and hickory are all fine, but be sure to avoid yew, sassafras and oleander) and remember that hardwood will make a more solid finished container than softwood.

Scrape off a top layer of bark so that you have a flat surface. The larger the area of this flat surface, the bigger the container you can make.

Carefully scoop (not with your hands!) some of the embers from the fire and arrange them in a circle (or whatever shape you want your vessel to be) on the flat surface of your wood. Blow gently on the embers. You'll find that they will burn downwards into the wood and will start to form the shape of a bowl. After an hour or two (depending on how hard the wood is), you should have a useable vessel.

Now extinguish the embers and empty the bowl. You'll need to use a sharp stone to scrape away the charred wood from the inside, then a rounded pebble to smooth it off.

Since our improvised pot is made of wood, we can't just stick it on the fire. But we *can* cook with it, using the time-honoured method of hot rocks. Find yourself a stash of small, rounded rocks – bigger than marbles, smaller than tennis balls. Try to avoid rocks from stream beds as these can explode when you put them in a fire.

(This happens because, if the rock is slightly porous, water gets into its fissures. When the water is heated, it turns into steam and expands, forcing the rock apart suddenly and very violently.) Place the rocks in the fire and heat them through for a couple of hours. Now, fill your improvised pot with handfuls of wild greens or whatever else it is you want to boil – you can cook animal parts and fish this way too – and fresh water. Drop one of the stones into the water, then keep adding stones until the water bubbles. Soon the boiling will start to die down. Remove the original rocks and replace with new ones from the fire. Keep doing this until your food is cooked.

PLANTS AND POISON

The natural world can provide you with food in abundance. But . . . it can also kill you.

This makes sense when you think about it. Most wild creatures in the animal kingdom have developed self-defence mechanisms – sharp teeth, claws, venom or even just the ability to run away fast. Wild plants don't have these advantages, so over time many of them have evolved a different way to fight back against predators that would like to eat them: by being poisonous.

If you're going to rely on the natural world to provide your food, it's as important to be able to recognize toxic plants as it is to identify edible ones. The pages that follow contain some of the more common poisonous plants and berries. ***This is by no means an exhaustive list.*** Whether you're out in the field or in a survival situation, it's very important that you don't eat any part of any plant unless you have positively identified it as being safe. Because some of the plants nature throws up aren't just toxic. They're deadly.

Before I describe some specific poisonous plants, I want to tell you about two poisons that are very common in the natural world and which you can learn to recognize and avoid. They are hydrocyanic (or prussic) acid and oxalic acid.

HYDROCYANIC ACID

This is a potent, water-soluble poison. It is very distinctive, and is normally described as having the smell and taste of bitter almonds. There is a plant called cherry laurel, with large, glossy leaves, which people often grow in their gardens. If you can find it, crush one of the leaves and smell it. That's the smell of hydrocyanic acid. If you come across it in the wild, avoid. It's poisonous even in small quantities. In large quantities it can be fatal.

OXALIC ACID

This is also very common. Rhubarb leaves, for example, are high in oxalic acid, which is why they are poisonous to eat whereas the stems are edible. A plant that is high in oxalic acid will give a sharp, burning sensation, and maybe some swelling, on the skin or tongue. See the Universal Edibility Test on page 57 to learn how to try this reasonably safely, and if you feel this swelling or burning, ditch the plant. Oxalic acid can be deadly in large quantities.

In general, you should avoid any plant that tastes very bitter, very acidic or very hot. These are danger signals.

SOME VERY POISONOUS PLANTS

DEADLY NIGHTSHADE

The Latin name of deadly nightshade is *belladonna*, which in Italian means 'beautiful lady'; it is thought to have come about because Venetian women used to make eyedrops from the deadly nightshade plant in order to make their pupils dilate attractively. This is a very, very bad idea. All parts of the deadly nightshade plant are poisonous. It has long, oval green leaves, purple, bell-shaped flowers and attractive, shiny black berries. The root of the plant is the most toxic part, but it's the berries that are most dangerous because they look so tempting – especially to children – and have a slightly sweet taste. In fact, eating a single berry could kill a small child. The toxins attack the nervous system, and the symptoms include convulsions, delirium, hallucinations, comas and, of course, death.

FOXGLOVES

Foxgloves are very distinctive, very common in the wild, and very poisonous. They have colourful, tube-shaped flowers, most often pink, purple or yellow. All parts of the plant are extremely poisonous: it can cause nausea, vomiting, hallucinations and heart failure.

YEW TREES

Almost every part of the yew tree is poisonous to humans, the exception being the flesh just around the seed of its red berries – though please don't try to prove that, because the seed itself is very toxic. The leaves are the most poisonous part of the tree and can cause heart failure or death when

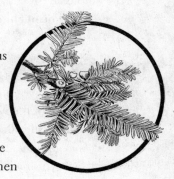

ingested. Symptoms include muscle tremors, breathing problems and convulsions. Sometimes, though, there are no symptoms at all. You feel fine for a few hours, then you die.

BUTTERCUPS

They might look pretty, but the entire plant is toxic. Probably not fatal in small amounts – and because they taste horrible you probably wouldn't try to eat more than one or two – but even a small amount can cause abdominal inflammation, diarrhoea and bloody urine. Not one for the pot, then.

LILY OF THE VALLEY

Talking of pretty flowers with a sting in their tail, lily of the valley is thought to contain more than forty different types of poison. All parts of it – including the red berries, which might look like food if you don't know what you're dealing with – are toxic. Lily of the valley causes abdominal cramps, violent vomiting, irregular heart rates and in some cases total heart failure.

OLEANDER

Oleander is very common in warm climates. It grows in the wild and is also a very common cultivated plant. Every part of it is poisonous, and you can die from swallowing a single leaf. The toxins in oleander hit your nervous system, your cardiovascular system and your digestive system. Triple whammy. Fortunately it tastes disgusting.

WHITE MANGROVE

A good example of why you should be extra careful in the tropics, even though the vegetation looks lush and distinctly edible. This tropical plant, found in Africa, Indonesia and Australasia, has beautiful white flowers and berries. The sap will make your skin break out in blisters. Get it in your eyes and it will blind you.

THE UNIVERSAL EDIBILITY TEST

The list of poisonous plants goes on and on. I hope that being able to identify these eight common ones will make the point that it really is a *very* bad idea to eat anything that you haven't positively identified as being edible first. At best, toxic plants can put you on your back; at worst they can put you in your grave. So what I am about to explain next must be understood with that caveat in mind.

The Universal Edibility Test applies to plants, not fungi. As you'll learn later in the book (see Chapter 3) there are mushrooms out there that taste delicious and can give you no instant reaction. But if you eat just the tiniest bit, you're a goner.

Sometimes, in a survival situation, you might come across a plant that you're unable to identify but which could mean the difference between life and death. In that situation, there is a way of reducing your likelihood of ingesting anything toxic. This is called the Universal Edibility Test. *It's not foolproof, and it's not to be used for fun.* Please don't imagine that knowledge of the Universal Edibility Test is *any* substitute for being able positively to identify edible plants.

Only use this test if your survival depends on being able to eat a plant you can't recognize.

1. **DISSECTION.** Some parts of a plant might be toxic, other parts might be edible. There's no point establishing that the leaves of a plant are safe to eat, then gobbling down the petals. So first, dissect the plant you want to test. Make sure you perform the following steps for each part you might want to eat. The flower, for example, needs to be divided into petals, sepals, stigma and stamen.

2. **INSPECTION.** Carefully examine the plant you want to test. You need to be sure that no part of it is rotting, and that no worms or other insects have taken up residence. Smell the plant to check that there is no hint of the bitter almond smell that indicates the presence of hydrocyanic acid (see page 54).

3. **PERFORM A SKIN-CONTACT TEST.** If any part of the plant causes irritation to your skin, you probably don't want it in your gut. Squeeze a bit of juice from the plant on to a patch of tender skin – the inside of your wrist, for example, or your upper arm just by your armpit. Now leave it for several hours and see if there is any reaction. If you get a burning sensation, bumps or redness, discard the plant and don't eat it. If there's no reaction, go on to step 4.

4. **CHECK FOR A REACTION ON YOUR LIPS AND TONGUE.** Do them in that order. Squeeze some of the plant's juice on to the corner of your mouth. Wait for a couple of minutes. If you experience any burning or irritation, rinse your mouth out with fresh, clean water and throw the plant away (it may taste nasty, of course, but that's a different thing – we're testing for edibility, not deliciousness). If you don't feel any discomfort, do the same to your lips, then the tip of your tongue, then the area under your tongue. At each stage, if you feel any burning or discomfort, you should consider the plant inedible.

5. **SWALLOWING.** If you've got this far, you're ready to ingest some of the plant. Don't go crazy. Swallow only a *tiny* amount. Now wait for several hours – five at least, eight if you can. Don't eat anything else, because you want to make sure that if you have a reaction, you know what's causing it. If you feel at all nauseous, or experience stomach cramps or dizziness or any of the warning signs of illness, you need to drink a lot of water and then make yourself sick by sticking your fingers down your throat (see below for what to do if you think you've poisoned yourself). If there is no reaction, you can consider it *reasonably* safe to eat the plant in slightly larger quantities.

WHAT TO DO IF YOU THINK YOU OR YOUR BUDDIES HAVE POISONED YOURSELVES

Mistakes happen, even if we do our best to avoid them. If you or anybody you're in the field with exhibits any of the following symptoms, there's a chance you or they might have been poisoned:

- sickness
- dizziness
- palpitations
- difficulty breathing
- seizures
- drowsiness
- loss of consciousness

If you suspect poisoning, you should try to get medical help immediately. If possible, call for an ambulance. It will be a great help to the medical staff if you can give them a sample of the suspected poison, tell them how long ago it was eaten and in what kind of quantity. While you're waiting for medical help to arrive, the patient should lie in the recovery position (on their side with the upper leg slightly raised so the knee juts forward, the upper arm bent at the elbow and the hand placed underneath the head to support it; while the lower arm is extended at right angles to the body and either bent at the elbow so that the hand is pointing towards the head, or just left extended straight). The head should face slightly downwards so that if the patient is sick the vomit can leave their mouth without choking or suffocating them. If they start to lose consciousness, do your very best to keep them awake.

If you can't call for medical help, your first thought should be to get to an area where help is available. In the meantime, there are some things you can do to help yourself or your patient. First, drink plenty of warm water. Try to make yourself vomit by putting your fingers down the back of your throat. If you've had a campfire, swallowing a little cold charcoal may help you to vomit. Charcoal can also stop your blood absorbing certain poisons. Mixing white ash into some fresh water creates a field medicine that can ease stomach cramps and can sometimes act as an antidote to hydrocyanic acid poisoning (see page 54).

SURVIVAL DRINKS

The best survival drink in the world is water. But sometimes, in the field, we need a drink that's going to give us a bit more of a morale boost. Maybe it's just the Englishman in me, but I don't reckon anything raises the spirits better than a cup of hot tea! If you've got a dry teabag or two in your pack, that's great, but it would be a shame to ignore all the great ingredients nature has lined up for us to make our own hot drinks. So here are three of my favourite teas that you can make in the wild.

PINE NEEDLE TEA

Pine trees are native to most of the northern hemisphere, and have been introduced into most temperate and subtropical regions around the world. So a taste for pine needle tea will keep you refreshed across vast areas of our planet! But this tea isn't only a great drink because the pine tree is so common. The needles themselves are one of nature's super-foods. They contain massive amounts of Vitamins A and C – a cup of pine needle tea will give you substantially more Vitamin C than a glass of orange juice, and Native Americans used to use it to prevent scurvy. The tea is a natural decongestant and expectorant, and has antiseptic properties. Also, it tastes good – mild and citrusy. In fact, I tend to chew on the young buds as I am hiking through pine terrain – like nature's candy!

To make pine needle tea, collect a handful of pine needles – the younger the better. Strip them from the branch, then chop them up into small bits about a centimetre long. Bring a cup of water to the boil and remove it from the heat. Add the pine needles, then let it infuse for about 10 minutes, or until the needles have settled to the bottom of the cup.

There are some poisonous pines that you shouldn't use to make tea. One is the ponderosa pine, native to North America. The other is the Norfolk Island pine, or Australian pine, common across the South Pacific. Yew trees are pine-like conifers and very poisonous: beware and avoid.

Another way to utilize the pine tree for food is to dig up its young roots. They have the texture of celery and can be a great source of calories in an emergency. You can also gather nuts from the cones of pine trees, but this can be a fiddly and time-consuming job and some varieties produce nuts that are too small to bother with. Make sure you're not expending more energy than you're likely to gain from the nuts.

BIRCH TEA

As we've already learned, the birch tree is one of the most useful trees you can come across in a survival situation. The bark can be used to make all sorts of waterproof vessels, from bowls (see page 24) to shoes to canoes! The bark also makes brilliant tinder that will take a spark directly, not to mention a fantastic drink.

To make birch tea, harvest some small twigs from the end of a branch, or alternatively scrape some bark away from new growth. Add the birch to a cup of boiling water and infuse for 10 minutes.

The birch tree also has a hidden secret: the sap. You can tap this off in the early spring to harvest some delicious, safe-to-drink liquid, full of sugars, minerals and Vitamin C – think of it as nature's energy drink. If you use this sap in place of water for your tea, the results will be awesome. Here's how to get at it.

Start off by making a cut in the tree – an upward slice into the bark is best. Now insert something to act as a tap – a short length of

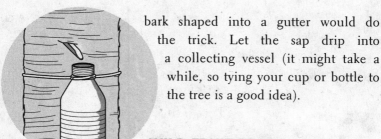

bark shaped into a gutter would do the trick. Let the sap drip into a collecting vessel (it might take a while, so tying your cup or bottle to the tree is a good idea).

WILD FRUIT TEAS

Most edible berries have leaves that can be turned into tea. Blackberry and wild raspberry leaves are fantastic, as are blueberry. Simply steep the leaves in boiling water for 10 minutes. You can do the same with lots of herbs – if you come across a patch of wild mint, you have the basis for one of the best, most refreshing teas in the world.

3

FUNGI

'Mushrooms are like men – the bad
most closely counterfeit the good.'
PAUL GAVARNI

If you've ever come across a patch of mushrooms when walking through a cool, damp forest or a wide expanse of open grassland, you'll know how abundant they can be. Knowing how to identify certain edible mushrooms is not only a pleasure, it's also a great string to your bow when it comes to wild food survival.

However, certain mushrooms are just about the most poisonous things the natural larder can provide you with.

They say that you can be an old mushroom hunter, or a bold mushroom hunter, but you can't be an old, bold mushroom hunter. It's worth remembering that. Getting things wrong with respect to the mushroom kingdom is potentially far more fatal than getting things wrong with the plant kingdom. Certain, quite common, mushrooms are deadly poisonous. There's no antidote, and in many cases nothing a doctor can do for you once you've ingested certain types of fungus. You're more likely to need the services of an undertaker instead: people die from mushroom poisoning, all over the world, every year.

In this chapter we're going to learn about what mushrooms are, which will help us understand where and when to find them. We're then going to look at some of the most toxic mushrooms you're likely to come across: this is important knowledge that can save your life whether you're in a survival situation or simply out in the field with your friends. Only then will we look at some edible mushrooms: I've chosen a handful that are fairly common all over the world, and which are pretty difficult to confuse with anything else. But even then you should only think about consuming them if you're *absolutely* sure you've got your identification right.

WHAT ARE MUSHROOMS, AND HOW DO WE IDENTIFY THEM?

Have you ever noticed how mushrooms often appear in the same place, year after year, sometimes in large circles that children like to call 'fairy rings'? This gives us a clue as to the true nature of the mushroom.

When you see a mushroom growing out of the ground, or on a tree, or on a dead branch, you're not actually looking at the whole organism. In fact, you're looking at the reproductive organs, like the fruit on a tree. The main part of the organism is underground (or in the wood of the tree). It's called the mycelium and is made up of thousands of microscopically thin strands. The mycelium sends up its fruit – what we call the mushroom – in order to disperse millions upon millions of tiny spores and so reproduce.

Because everything that the mushroom needs to grow is stored in the mycelium, mushrooms can pop up extremely quickly. One day you might see an empty field, the next it could be stuffed full of emerging field mushrooms.

There are so many thousands of different types of mushroom that you could (and many people do) spend a lifetime learning about them and still not know it all. There are hundreds of books and field guides on the subject, and the internet is a rich source of information. But really, you can never hope to learn how to gather wild mushrooms successfully simply from reading a book. You need to get out there and start rummaging through the forests, preferably with somebody who has extensive experience of mushroom-hunting. Most importantly, you need to know what to look for in a mushroom if you want to have any hope of identifying it. There are three main criteria you have to consider: its season, its habitat and its appearance.

SEASON

Most people think of autumn as being mushroom season, but there's a bit more to it than that. Sure, many mushrooms emerge in the autumn, but some can keep going through a mild winter and others come out in the spring. Knowing the seasons of different mushrooms can be invaluable when identifying them.

HABITAT

Some mushrooms grow only near beech trees, others grow only near oak. Some pop up individually, others in clusters. Make a careful note of where you have found your mushroom and you'll have a much better chance of identifying it correctly.

APPEARANCE

You need to become practised at examining various parts of the mushroom. They are:

- The cap – its size, its shape, its texture and whether the skin can be peeled.
- The stem – its size, shape and texture. Does it have a ring around it like a collar or a skirt? Does it have a bulbous 'bag' at the base?
- The gills or tubes beneath the cap – their colour, how far apart they are, their texture.
- The flesh – is it rubbery or crumbly? Does it have a particular smell? Does it change colour when cut?

This might sound like a lot to remember, but the only way positively to identify a mushroom is by examining every part of it carefully.

WHERE TO FIND MUSHROOMS

We've already said that mushrooms tend to be found on open grassland or the forest floor. But we can do a bit better than that.

Mushrooms don't like to compete with long grass, so as far as open grassland is concerned, they prefer areas where animals graze.

Forests are better hunting grounds, but you need to bear in mind that mushrooms don't much like competition. If there is a lot of thick undergrowth, the fungi might struggle to push through. It's also worth remembering that the mycelium might not feel the need to reproduce if it exists in very fertile ground. This is why you might have more luck close to well-worn pathways or at the edge of the forest – anywhere that the terrain changes from one type to another.

SOME OLD WIVES' TALES TO IGNORE

Sometimes, when you're out in the wild, it's worth remembering some of the old wives' tales that have been passed down from generation to generation. A lot of weather lore, for example, can give you a fairly accurate indication of what the elements have in store for you.

Not so with mushrooms.

- Don't listen to anyone who tells you that a mushroom is safe to eat if you can peel the cap (you can peel the cap of the most poisonous mushroom in the world).
- Don't listen to anyone who tells you that a mushroom is safe to eat because it's growing on wood (simply not true).
- Don't listen to anyone who tells you that if you see an animal eating a mushroom, it's safe for humans (unless you want to end up dead). Or that cooking a poisonous mushroom renders it edible (it doesn't, except in a handful of cases).

There are *no* shortcuts to identifying mushrooms. The *only* way to tell if a mushroom is safe to eat is by positively identifying it with a reliable field guide. The pages that follow will help you with some of the common ones. But I urge you to use as many different books and resources as possible to learn about identifying mushrooms. Because trust me: this is an area of wild food survival where you *really* don't want to make any mistakes.

MUSHROOM NAMES

There are thousands of different types of mushroom. Some of them have very distinctive common names, which can be really useful – you're probably not in much doubt about whether you should eat something called a Death Cap, or a Sickener, or Poison Pies. Most experienced mycologists, however, prefer to use the Latin names because individual mushrooms sometimes have several common names, so confusion can arise where you really don't want it. The Latin names are specific, and there's only one per mushroom. Here, I've given you both names.

SOME VERY POISONOUS MUSHROOMS

These are not all the poisonous mushrooms you're likely to encounter. Not by a mile. But they are all pretty common in lots of places around the world and you'd do well to learn how to identify them. Not because you're going to go ahead and eat any old mushroom that *isn't* on this list (at least I hope you're not), but because studying these mushrooms will set you on the right path to becoming a more experienced mycologist. And you might just be in a position to stop one of your buddies on the trail making a deadly mistake.

These poisonous mushrooms principally grow in Europe,

North America and Australia, but look out for them in all temperate and subtropical regions – wherever mushrooms grow, you need to have your eyes open for the bad guys.

DEATH CAP
(*Amanita phalloides*)

Where it grows: Europe, North America, Australasia.

If you mistakenly pick and eat *Amanita phalloides*, the chances are you won't live to regret it. If you eat the Death Cap, it will probably kill you. Untreated, it is deadly in nearly 90 per cent of cases. Even if you manage to get to hospital and doctors throw everything they have at it, it's fatal in about 20 per cent of cases.

And it doesn't just kill you. It kills you nastily.

If you were to eat a Death Cap, you'd probably find it extremely delicious – it is, by all accounts, very mushroomy. And for 8–12 hours after that, you'd probably feel fine.

Then you wouldn't.

Your first symptoms would include massive stomach cramps, violent vomiting and explosive diarrhoea. As a result of this, you'd most likely lose a lot of fluid, so your blood pressure would drop, your pulse would increase and you'd very probably go into shock. This would last somewhere between 12 and 24 hours.

After which you would suddenly recover. At least, you'd *think* you'd recovered. But you wouldn't have. Because over the next few hours, the toxin amanitin would be getting to work on your liver. You'd start to show all the signs of liver failure: jaundice, intestinal bleeding, pain, hallucinations . . . The poison would be turning your liver to jelly. Once that happens, there's nothing anyone can do for you. You'd slip into a coma and die shortly afterwards.

Scared? Good. The Death Cap is one of nature's bad guys. Here's how to identify it.

Habitat: oak, beech and general mixed woods. Grows in small, widely spaced groups of about six.

Cap: 5–15cm across, convex or flattened, smooth, green or yellow (though sometimes almost white) with a pale edge, sometimes slightly sticky. Look out for faint fibres radiating out from the centre.

Stem: 6–15cm long, 1–2cm wide, bulbous at the bottom. The bulbous part is surrounded by a sac-like 'volva'; the stem has a large, collar-like ring with a skirt hanging down.

Gills: white, not touching the stem.

Flesh: white, with a pungent, sickly smell.

WARNING

The young Death Cap can easily be mistaken for an ordinary button mushroom.

THE DESTROYING ANGEL
(*Amanita virosa*)

Where it grows: Europe, but there are similar varieties, also commonly called Destroying Angel, found all over the world.

You may have noticed that the Latin name for this mushroom includes the word *Amanita*, just like the Death Cap. This means they're both from the *Amanita* genus, which includes many poisonous mushrooms.

Amanita virosa is a pretty-looking mushroom. That's why they call it an angel. It's also deadly – just as deadly as the Death Cap – which is why they call it the *Destroying* Angel. There's no antidote to its poison and, like the Death Cap, the Destroying Angel causes massive organ failure. If you mistakenly put one of these in the pot, your only real chance of survival will be a liver transplant. Not easy in the field.

Habitat: general mixed woodlands; also found in fields and on road-sides.
Cap: 5–12cm across. Starts out conical, becoming flatter as the mushroom ages. White, smooth.
Stem: 9–12cm in length, 1–2cm across, swollen at the base, which is surrounded by a sac-like 'volva'. White, sometimes shaggy, with a white ring at top of stem.
Gills: white, crowded.
Flesh: firm and white with a sickly smell.

FLY AGARIC
(*Amanita muscaria*)

Where it grows: native to most temperate zones of the northern hemisphere, it has also been introduced to the southern hemisphere and is found in many parts of the world.

This is the mushroom that pixies sit on in children's story books. You know the one – a red cap (sometimes orange), with little white spots. It's extremely common, and because of its distinctive colour it tends to jump out at you when better-camouflaged mushrooms are more difficult to spot.

Don't eat it. The Fly Agaric is nowhere near as poisonous as the Death Cap or the Destroying Angel – although deaths have been attributed to it – but it has some very nasty effects: drowsiness, difficulty in speaking, confusion, spasms, cramps, tremors, hallucinations and deep coma. Some people eat Fly Agaric for fun and in certain cultures it is prized for these effects. But you need your wits about you in the field. The last thing you want is to have your mental sharpness compromised. It's called Fly Agaric because people used to use its sap to stupefy flies. Don't get stupefied yourself.

Habitat: forest floors.
Cap: 8–20cm. Starts out conical, becomes flatter as the mushroom ages. Bright red or orange, covered with white spots.
Stem: 8–20cm in length, 1–2cm across. White, sometimes shaggy, with a bulbous base and a simple white ring at top of stem.
Gills: white.
Flesh: white, sometimes tinged with red or yellow.

PANTHER CAP
(*Amanita pantherina*)
Where it grows: North America and Europe.

Another deadly *Amanita*. It contains the same types of toxins as the Fly Agaric, but in much greater quantities. The Panther Cap is a good example of a mushroom that can easily be confused with an edible mushroom, the Blusher. If you were to put the two side by side, you'd see the difference. But you'd also notice the similarities, which are a very good reminder that you need to be extremely sure of your identification before you risk eating a wild mushroom. The Panther Cap is seriously poisonous, causes sickness and hallucinations, and can easily kill you, especially if you have a weak heart.

Habitat: forests, especially near beech trees.
Cap: 6–12cm across. Starts out conical, becomes flatter as the mushroom ages. Chocolate-brown, covered in small white spots.
Stem: 6–13cm in length, 1–2cm across. White, with a bulbous base wrapped in a white 'volva' and a white ring at top of stem.
Gills: white, not touching the stem.
Flesh: white and mild-smelling.

FOOL'S FUNNEL
(*Clitocybe rivulosa*)

Where it grows: Europe and North America.

The Fool's Funnel is sometimes called the Sweating Mushroom, because that's what its main toxin makes you do. Before it shuts down your respiratory system, causes cardiac failure and kills you, that is. Another name for it is the False Champignon, and it has a nasty tendency to crop up near the edible and popular Fairy Ring Champignon (*Marasmius oreades*), although it does look quite different to that. Fool's Funnel frequently appears on grassland, and is one very good reason why you should closely monitor children around 'fairy rings'. Such rings might be harmless, but they could be deadly.

Habitat: grassland, sandy soil, roadsides. Often appears in circles.
Cap: 2–5cm across. Convex around the edge with a small depression in the centre. Dirty white in colour. Sometimes develops concentric rings.
Stem: 2–4cm in length, 4–10mm across. Dirty white in colour (like the cap), furry at the base. No ring.
Gills: crowded and off-white.
Flesh: white to off-white and sweet-smelling.

So there you go. Five mushrooms that can do some pretty horrid things to your insides, and which you should never put anywhere near your mouth. Scared? Good. Some mushrooms are at least as toxic as venomous snakes.

But I don't want to put you off entirely. Once you know what you are doing, hunting mushrooms can be a great outdoor activity. More importantly, the ability to identify certain edible species can be invaluable in a survival situation. The following are five fungi that are relatively easy to recognize. Once you've identified these in the wild, I almost guarantee you'll want to find out more about the fascinating world of mushrooms.

SOME EDIBLE MUSHROOMS

First, some dos and don'ts.

1. Never eat a mushroom if you haven't positively identified it. Use more than one source to do this.

2. A mushroom might be edible, but if it's past its best it can still cause you harm. If it looks or smells suspicious, is discoloured in any way, or shows any sign of maggot infestation, chuck it away.

3. Although some fungi are edible raw, many are toxic unless cooked, so it's a good idea to cook (ideally boil) any mushroom you've found in the wild. (But remember, cooking does *not* make *all* poisonous mushrooms safe.)

BEEFSTEAK FUNGUS
(*Fistulina hepatica*)

Where it grows: Europe, North America, North Africa, Australia.

The beefsteak fungus is a pretty weird-looking thing. It certainly doesn't look like your regular mushroom. It grows on living trees – normally oak or chestnut. When it's young it looks like a large tongue (its French name is *langue de boeuf*). As it gets older, it starts to resemble a liver (the name *hepatica* is Latin for liver). If you cut into the flesh you'll see that it looks like a piece of fatty steak. Even stranger, it'll look as if it's bleeding because it will exude a large amount of red fluid. Beefsteak fungus has a rather sour, acidic taste – you'll probably want to soak it first, then stew it well – it needs quite a lot of cooking before it's tender enough to eat, unless it's very young.

Habitat: live oak and sweet chestnut trees.
Cap: 8–25cm across, dark red on top, pinkish below.
Stem: none.
Gills: cream-coloured when the mushroom is young, and easy to separate from one another.
Flesh: red with white marbling. Exudes dark red liquid when cut.

CHANTERELLE
(*Cantharellus cibarius*)

Where it grows: Europe, North America, Asia, Africa.

If you stumble across a patch of chanterelles, you're lucky. They're not only edible, they're delicious. They're also pretty easy to identify. They're egg-yellow, have a distinctive funnel shape and smell slightly of apricots. Best eaten fried, but as a survival food can be boiled for 10 minutes before eating. There is an impostor called the False Chanterelle (*Hygrophoropsis aurantiaca*), which is even more common than the chanterelle. Fortunately it's not poisonous, but it doesn't taste nearly so good. The *poisonous* mushroom that is easiest to confuse with the chanterelle is the Jack O'Lantern (*Omphalotus illudens*) – so called because its gills glow in the dark. It causes stomach cramps and vomiting in humans, but is relatively rare.

Habitat: woodland, especially under oak, beech and pine trees.
Cap: 3–10cm across. Funnel-shaped, with the edge rolling downwards. Egg-yellow in colour.
Stem: 3–8cm in length, 0.5–1.5cm across, of a similar colour to the cap.
Gills: more like grooves than gills, the narrow veins fork into two and continue down the stem.
Flesh: yellow, but not as dark as the surface of the cap.

GIANT PUFFBALL
(*Calvatia gigantea*)

Where it grows: very common in the UK; also found in Europe and North America.

Giant Puffballs are awesome. They are totally distinctive – big white globes with no gills, that can grow to the size of a football – and extremely edible. If you find one, chances are you'll find a batch of them in a ring. As tempting as it might be, take only what you need. A puffball's flesh should be white all the way through if it's to be edible. If it starts to go brown or even green, it means it's getting ready to release its spores and shouldn't be used for the pot. To cook it, peel off the thick skin, then cut it into slices or chunks. It's fantastic fried over the campfire with a little oil, salt and whatever other flavourings you have to hand. You can also boil it, though it will lose some of its texture that way. However you cook it, a good slab of puffball can make you a substantial meal when you're hungry in the field.

A small word of warning: Giant Puffballs are almost impossible to mistake for anything else once they've grown to their full size, but baby specimens can be confused with baby Death Caps or Destroying Angels. If you find yourself wanting to pick a 'mini' Giant Puffball, think twice.

Habitat: grassland and among old nettle beds; occasionally in forests if the soil is very rich.
Cap: varies widely in size; can grow up to 80cm in diameter but more often around 30cm. Pure white skin.
Stem: none.
Gills: none.
Flesh: white and spongy, turning brown/green as the mushroom ages.

CHICKEN OF THE WOODS
(*Laetiporus sulphureus*)

Where it grows: Europe and North America.

Another fungus that is really pretty easy to identify. Chicken of the Woods grows on trees – mostly oak – and has a distinctive bright yellow and orange colouring. It gets its name from the fact that it supposedly looks like a chicken's foot, which I suppose it does if you use your imagination a bit! Some people have an allergic reaction to this fungus, so make sure you cook it well and eat only small portions to begin with. It has a decent taste – slightly sour, but good and mushroomy.

You might sometimes find Chicken of the Woods growing on yew trees. If you've read the section on yew (page 55) you'll know that it's not a good tree to get involved with. Don't eat any fungus you find growing from it.

Habitat: live trees, mainly oak, sometimes cherry, sweet chestnut, willow and yew – avoid those on yew.
Cap: semi-circular/fan-shaped, 10–40cm across. Bright yellow and orange in colour, tiny pores on the underside.
Stem: none.
Gills: none.
Flesh: the young flesh releases a yellow juice when squeezed, but it becomes crumbly with age.

CAULIFLOWER FUNGUS
(*Sparassis crispa*)

Where it grows: Europe and North America, cultivated in Japan.

Here's another one it's really hard to mistake. The cauliflower fungus looks like a spongy cauliflower head. You find it growing at the bottom of conifers, and nothing else really looks like it. Because of its shape, you might find that all manner of creepy-crawlies take up residence inside it. You can get rid of them by immersing the cauliflower fungus in (purified) water and waiting for them all to come slithering out. You can cook a cauliflower fungus pretty much any way you want – fried, boiled, stewed – and it tastes great. Try sticking slices of it into one of the campfire stews on page 36.

Habitat: at the base of, or near, conifers.

Stem: none.

Gills: none.

Flesh: looks like a – you guessed it – cauliflower with lots of 'crisped' lobes, 20–50cm across, creamy colour, sometimes turning darker at the edges, sweet smell.

Wild 'Pizza'

To make this, you'll need some wholewheat flour, water and salt, plus whatever wild ingredients you've managed to get your hands on. Any of the edible mushrooms above would be good, as would some boiled nettles, or fish, or cooked seafood (see pages 108–25).

Make a soft dough by mixing the flour and water in a ratio of about 2:1. Add a good sprinkling of salt and knead the dough till it's well combined. Now press or roll it into a disc (you can use your water bottle, if you have one, as a rolling pin).

Get a small frying pan hot over the embers of a (not too fierce) fire and add your dough. Cook it till the underside is brown, which will take up to 5 minutes, depending on how hot your fire is. Flip it, then sprinkle your topping ingredients over the cooked side which is now face up. If possible, cover your pizza with tin foil or with the lid of another pot so that the toppings can heat through.

Return the pizza to the fire and cook until the underside is brown and the toppings are hot – again, about 5 minutes.

You could add anything to your wild pizza, of course. If you've harvested some insects (and in Chapter 11 I'll tell you all about that), they'd make an excellent and nutritious addition. I can tell you from experience, including some roasted crickets adds an awesome crunch!

TWO COOL THINGS YOU CAN DO WITH MUSHROOMS THAT DON'T INVOLVE EATING THEM!

There are two fungi that are useful in the wild for reasons other than keeping your belly full. The first is known as the Tinder Fungus. The second is called the Razor Strop.

TINDER FUNGUS (*Fomes fomentarius*)

The Tinder Fungus is inedible, but it has been used for thousands of years by indigenous people because part of its flesh makes an excellent tinder for firelighting. The fungus itself grows on birch trees and has a hard, grey-black outer crust. If you cut into it, underneath are the pores. The final, cork-like, orange layer is the 'trauma'. This is the part you want. Slice a piece of this off, let it dry out, then fluff it up a bit with the flat of your knife. Drop a spark on to it. The fungus will burn slow and hot. It won't produce any flame, so you'll need to add some dry grass if you're going to use it to light a fire. By itself, a decent-sized slice of fungus will smoulder for quite a long time. For this reason, it can be used to 'save' embers, or to transport fire.

RAZOR STROP (*Piptoporus betulinus*)

Like the Tinder Fungus, the Razor Strop grows on birch trees. It's also inedible, but indigenous peoples have long used its leathery flesh to sharpen their knives. Cut a slice out of the fungus and let it dry (the size of the slice will depend on the size of the knife you want to sharpen). Your dried piece of fungus can now act as a strop. Use it to hone the blade of your knife by swiping the blade along the strop (imagine spreading butter back and forth over a piece of bread). Stropping a knife in this way will make a sharp edge even keener.

4

FISHING

Fish are great to get your hands on in a survival situation. They taste awesome and they're *incredibly* good for you, as they contain a good mix of all the right proteins, fats and vitamins. There are many stories of men and women adrift at sea surviving for months on nothing more than the fish they could catch. So whether you're just having some fun with your buddies, or you think you might end up in a situation where finding your own food is a matter of life and death, learning how to catch fish in the wild is an essential skill.

First, though, I've got some bad news: fishing ain't what it used to be.

In the days before our seas and rivers became polluted with the toxic by-products of modern industry, and fish stocks had not been decimated by over-fishing, aquatic life was far more abundant than it is now. It's thought that the sport of fly fishing – when you attract fish using a brightly coloured lure – was popularized by Victorian fishermen because using live or dead bait was seen as too easy and unsportsmanlike.

Unfortunately, our waters are not quite so well stocked as they once were. When we're in the wild and relying on fish for sustenance rather than sport, we need to do everything we can to stack the odds in our favour. And if we find ourselves – as we are likely to – without a tackle bag full of modern fishing gear, we're going to have to rely on the old ways to catch our supper.

But here's the good news: water is still often the best place to find food. The even better news is that, having been honed over centuries of civilization, wild fishing techniques can be brilliantly effective. You don't have to spend a fortune at the tackle shop. In fact, you don't have to spend anything. Our ancestors didn't have a whole collection of flash, modern fishing gear, and they didn't do so badly. You can even catch fish with your bare hands, as I have often done in small rivers in the wild – it's called fish tickling and, although it's illegal in the UK because it disturbs their habitat, it's still permitted in many wildernesses around the world. Make sure you check first, unless you're in a genuine survival situation.

What's more, there's something very satisfying about landing a fish using nothing more than materials you've foraged from the wild. You won't only catch the fish – you'll catch the fishing bug too!

A word of warning: different parts of the world have different rules about fishing. Although you're unlikely to encounter any problems in the Arctic or the Tropics, you will need a rod licence when fishing for freshwater fish, eels, salmon or sea trout anywhere in England or Wales and up to 6 miles out to sea – even if you're not using a standard fishing rod. There may be other local restrictions depending where you are. Remember to check, unless you're in a genuine survival situation.

First we'll look at where and when to catch your fish, then we'll look at *how* to catch them. Finally, we'll look at how to prepare them.

WHERE AND WHEN TO CATCH YOUR FISH

If you're fishing a river, before you look at the water, take a look at the weather. If the sun is out and it's hot, the fish are likely to be in the deeper, colder waters. On the other hand, if it's cold out they're more likely to be in shallower waters where they can catch some of the sun's rays. And whatever the weather, you're likely to find fish in the shelter of large rocks.

Evenings are often a good time to fish a river: insects hover just above the water and bring fish to the surface looking for their dinner. If you see ripples in the still waters, it's a good indication that fish are there.

If you've ever put your hand in a swimming pool, you may have noticed how the bit under the water looks as if it's coming off at a different angle. This is because light bends – or refracts – when it enters water. As a result of this effect, fish can see a lot more than you might expect when you're near the bank of a river. So, if you're holding an improvised rod and line on a river bank, crouch low so the fish don't see you and get spooked. Alternatively, fish while it's lightly raining, as the impact of the rain on the water breaks up the fish's view.

If you're fishing from the seashore, a lot depends on where you are and what you're trying to catch. The general rule is that you should fish a moving tide, but an even better rule is to find someone who knows the waters well and ask them for some good local advice. There's no substitute for this kind of knowledge.

IMPROVISED FISHING GEAR

Humans have been catching fish for food for thousands of years. They haven't always had the benefit of high-end fishing gear to make the job easier. It's worth remembering that if you find yourself by the water. Here are some ideas for how you can improvise your own fishing gear.

HOOKS

If you're in the field, it may be that you have some fish hooks with you. In fact, if you're putting together a survival pack it's a very good idea to include a couple of hooks and some good, sturdy fishing line. They hardly take up any space.

But if you haven't, don't worry – hooks and line are easy to improvise. Hooks can be fashioned out of all sorts of things – paperclips, pins and needles, hairpins, ring pulls from cans of fizzy drink (there aren't many places in the world that are free of litter, but you can turn this to your advantage), bits of bone, flint or even sea shells. Use your imagination!

In the absence of any of these, you can use the tools with which nature has supplied us. Here are two methods of making a perfectly decent fish hook, one using a splinter of wood, the other using a bramble thorn.

Wood hook

Find or cut a splinter of wood about 2cm in length. Sharpen both ends with your knife. If you like, you can fire-harden the sharp ends. This involves holding the wood close enough to a heat source to dry out any moisture without burning the actual wood. You then need to tie your line or cordage (see page 87) to the middle of the hook.

When you add your bait, use it to position the hook so that it lies parallel to the fishing line. This makes it easy for the fish to swallow, but once it's inside the gullet it will open out at right angles to the line and become almost impossible to dislodge. It will also kill the fish, so make sure you intend to eat anything you catch this way.

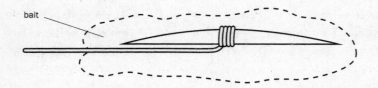

bait

Thorn hook

Brambles and other plants with sturdy thorns grow all over the place. Cut yourself a short length of the stem with a good, strong thorn attached, which will act as a hook. Tie it to your line so that the thorn hook is pointing upwards. You can even tie several hooks together, as shown.

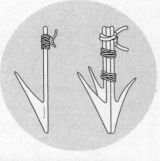

Don't make your hook too big. You can catch big fish with small hooks, but small fish will avoid big hooks if they're too much of a mouthful.

RODS

It's perfectly possible to fish without using a rod – just a line and a hook will do and, as you'll see below, sometimes using a rod is not the best option. But you'll be able to cast further and pull in your catch more easily if you do have one. Any branch will do, but a straight, slightly flexible one is the easiest to handle and will be less likely to snap if there's a bit of pull from the fish.

LINES

If you have some decent fishing line in your pack, so much the better. But in an emergency you can always make a line from any string you happen to have in your kit, from a shoelace, a length of unravelled thread from your clothes, or a length of sturdy, tall grass. It's also possible to make natural cordage from the materials nature provides you with. See Appendix A for details on how to do this.

FLOATS

Floats are essential to rod and line fishing because they allow us to dangle the bait at the right level below the water (see page 90). You can buy all sorts of fancy floats from tackle shops, but in reality all you need is something small that floats on water. A wine-bottle cork is good, and in many parts of the world people simply use scraps of low-density wood. Balsa wood floats particularly well – people have sailed across oceans in balsa-wood rafts before now.

Floats also have a second purpose. Sometimes it's hard to tell from the feel of your line if you've got a bite. But if a fish has taken your bait, it will probably pull the float beneath the surface.

You'll need a bigger float for sea fishing than for freshwater fishing, to cope with the currents and the waves.

WEIGHTS

Without a weight at the end of your line, your baited hook will probably just float on top of the water – fine if the fish are top-feeding, but no good if they're lurking on the river bed. Military survival kits often include some little spheres of lead shot which can be used to weight a line. It's a good idea to take some along with you if you think you might have to rely on the water for your food – they take up no space and weigh practically nothing, but they'll help you get your bait (see next page) under the water line. Sometimes you'll need a heavier weight. You can buy these, of course, but a stone tied to the end of your line will do just as well.

Once you've improvised your fishing gear, you'll need to tie it all together. Fishermen use all sorts of fancy knots and it might be worthwhile learning a few – there are loads of resources out there to help you do this. But in a survival situation, any decent, strong knot should do the trick. See the section on rod and line fishing on page 90 for some different configurations of your improvised rig.

BAIT

Dangling a bare hook into a patch of water is unlikely to catch you your supper. You need to persuade a fish that this is something worth going for. The best way to do this is to convince the fish that there's a meal in it for him.

You can dig up worms, of course. If they're plentiful, try to replace them often, because a wriggling worm will catch the attention of a fish more readily than a dead one. (And if there's a supply of worms, don't forget that they're a great food source in themselves if you don't land a fish – see page 211.) Alternatively, any scraps of food that you have about you – especially meat – will make good bait. If you catch a fish too small to make good eating, you can cut that up to make bait. Dead insects are another good option.

If the area of water you are fishing has an overhanging tree with berries on its branches, it's a good bet that fish have adapted to using them as a food source when they fall into the river. If you can gather these berries from the branches of any nearby trees, you can encourage fish to the surface by throwing them on to the water. Put a berry on to your hook and fish the same patch of water. Chances are you'll get a bite.

If you're able to gather insects – alive or dead – these are another great way to encourage fish into a feeding frenzy (and, like worms, they might also be a good food source for you – see Chapter 11). Scatter them on the patch of water you want to fish, just like the berries. In parts of the world where termite mounds are common, a good trick is to break off a piece of one of these mounds and suspend it from a branch overhanging the water. You'll get a constant drip-feed of insects into the water that should attract the local fish in their droves.

If you don't have anything that the fish would consider real food, it's worth remembering that they are attracted to colourful objects moving through the water that they might *mistake* for food. Professional anglers use 'spinners' – brightly coloured artificial baits that spin as you pull them through the water, attracting the fish's attention as the light glints off them. You could use a coin or a button, or even little bits of brightly coloured plastic. And at a push, you can improvise a lure from leaves or feathers.

TECHNIQUES FOR CATCHING FISH

ROD AND LINE

This is how you'll see most people fishing when they do it for sport, though if you really have to get food, it's not necessarily the best way. In order to fish with a rod and line, you'll need to make sure that you get your bait at the right level where the fish are feeding. In general it's hard to know whether the fish are surface-feeding, middle-feeding or bottom-feeding; you'll only *truly* find out by fishing at all different levels. So here are some combinations of hook, line, weight and floats that will allow you to do this. If you don't know where the fish are feeding, try all three at the same time. The more hooks you have in the water, the better your chance of catching a fish!

To fish near the top of the water, attach your float fairly close to the hook, as shown. Some small lead weights (you can get these from angling shops and attach as many as you need in a given

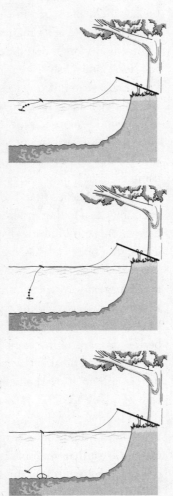

situation) will stop the hook and bait floating back up to the top of the water. Alternatively, a stone will do the job. The faster the current, the heavier the weight will need to be to keep the hook where you want it.

If you want to fish the middle of the water, extend the distance between the float and the hook accordingly.

To fish the bottom of the water, extend the distance between the hook and the float even further. Now attach a slightly heavier weight to the line, while leaving the hook and bait to float freely.

NIGHT LINES

First, a word of warning: night lines are illegal in many parts of the world. They should be used only in a genuine survival situation, and certainly never in waters where someone might unsuspectingly come across them.

The principle of a night line is to attach several baited hooks to a long piece of fishing line, weight the end, then cast the line overnight. There's a pretty good chance you'll end up with fish for breakfast.

You'll need to fix a sturdy weight to one end of the line to make sure that it remains taut. Anchor the other end firmly to something on the bank that isn't going to move. A tree trunk would be ideal, but failing that you can drive a peg deeply into the soil. Have several baited hooks attached to the line, as shown.

It's called a night line because you can leave it all night unattended then come back in the morning to see what you've caught. You could, of course, use this method during the day too, but only in a genuine survival situation.

GILLNETTING

This is the process of setting up a 'wall' of netting under the water. On a commercial scale it's illegal in many parts of the world because it's such an effective way of scooping up vast quantities of fish indiscriminately. Even small-scale gillnetting can be against the rules. But if you have a net and your survival depends upon it, this is a good way to gather much-needed food.

In order to set up a gill net, you'll need weights and floats. The weights keep the net anchored to the river or sea bed, while the floats keep the net upright. Any fish that hits the net will become tangled and caught. Don't use a gill net for too long – it's important not to take more fish than you need to eat.

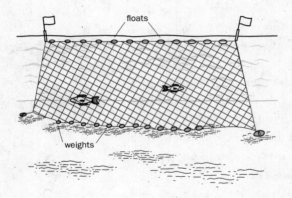

SPRING SNARE FOR FISHING

In Chapter 7 you'll learn how to make some different snare traps to catch small game. One of these is called the spring snare, and you can adapt it for fishing. See the picture on page 93. You need to fashion a trigger out of two pieces of wood by carving a hook at one end of each of them with your knife. One of these pieces of wood is staked into the ground. The other is attached to a springy sapling or overhanging branch and has your line, hook and bait attached to the other end. The hooks of the two pieces of wood are then hooked together, forming a sort of trigger. When a fish bites and gets caught on the fish hook, it will struggle. This will release the trigger, causing the sapling to straighten and pull the fish directly out of the water.

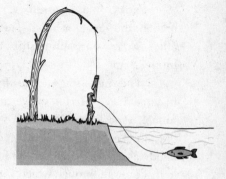

Make sure to check a spring snare regularly, especially in warm weather – out of the water, the fish can start to spoil very quickly.

FISH TRAPS

Fish traps can be extremely effective – so much so that even commercial fishermen use them, especially to catch lobsters and crabs. If you know the techniques involved, they can be a brilliant way of hunting for your supper.

I'm going to show you two clever ways of making fish traps in the wild: basket traps, and fixed traps that you create on the river bed or seashore itself.

Basket traps

The idea of a basket trap is to create a structure that a fish can easily swim into, but which it will find hard to swim out of. If you look at the picture on page 94, you'll see an inverted cone at the open end that serves this purpose.

To start with, you'll need seven (it's important that it's an odd number, because of the up-and-under weaving that you're going to do later on) straight, flexible strips of sapling, about 150cm long. Lay them on the ground and lash them tightly together at one end so that they make a neat circle. You can place them round an extra central stick to keep them in position if you like.

Now, find yourself a very flexible branch and create a hoop. Tie this to the other end of your sapling strips to make a funnel shape.

The next stage is to find some material that you can weave around the funnel. Vines are good. Sturdy grass will do the job. You need to weave it carefully and tightly along the length of the funnel. The idea is to make a wall through which your prey cannot escape. You'll find that when you've done about half of it, you can remove the hoop at the end, as the weaving will now hold the sapling strips in shape.

Now it's time to make the inverted funnel for the open end. To do this, take an odd number of short sticks and place them in the ground so that they fan out in a funnel shape. The small end should be just big enough for a fish to swim through, and the open end needs to have about the same circumference as the open end of your first funnel. So cut the length of your sticks accordingly.

Weave a wall round this second funnel in the same way. You're almost ready to go!

Now you need some bait – the bigger the better so that the smell really attracts the local fish population. Hang it inside the larger funnel, then fit the smaller funnel inside the larger one and tie the two funnels securely into place.

You're now ready to lay your trap. Tie it to a tree or a fixed point near the water, then throw it in. Leave the trap overnight. With any luck, you'll wake up to discover that a fish has found the lure of your bait irresistible, and has swum inside the trap and found it impossible to escape. Breakfast is served!

Fixed traps

These work on the same principle as basket traps: they allow fish to swim in, but make it difficult for them to swim out. You make a shape in the water, using rocks or sticks, that funnels any fish that happen to be swimming in that direction.

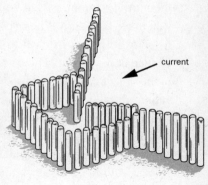

current

A fish trap such as this relies on the funnel facing the direction of the current, rather than on attracting fish using bait. You must make sure that the 'wall' has no gaps in it that are large enough to allow your prey to escape. And while a few fish *might* escape back through the funnel, this can be a very productive way of fishing.

SPEARFISHING

One of the oldest, most primitive ways of catching fish – but still effective if you know how to do it. You can make a spear by sharpening a good, sturdy stick. Even better, you can take a thick pole of bamboo and, with your knife, make four intersecting crosscuts into the end, each at 45 degrees to the other, so that you get eight prongs. Stick a smaller piece of bamboo into the centre of the cut end. This will make the prongs splay out a little. You can separate them further by wedging more small sticks between the cuts. Sharpen each of the prongs, then throw away the wedges and there you have it: a much more effective tool.

Alternatively, you can attach a sharp piece of flint or anything that will pierce the fish's skin and hold it fast. Be wary of attaching your own survival knife to a spear – it's an easy way of damaging it. See page 166 for a few more thoughts on how to make a good spear.

Spearfishing works best when there are plenty of fish and the water is fairly shallow – any higher than your waist and you won't really be able to do it. Try to position yourself so that you don't cast a shadow over the fish you are trying to catch, because this will spook them. And remember what we learned about the way light refracts in water (see page 85). When you look at a fish, it isn't actually where it seems to be – you need to aim slightly *below* the fish to have a chance of spearing it. The best technique is slowly to move the point of the spear as close to the fish as possible. Then, with a sudden, sharp action, pin the fish to the river bed. Once you've managed to spear your fish, reach into the water and grab it with the other hand – it will probably get away if you lift the spear.

You can use a bow and arrow to catch fish just as you can use a spear – I've successfully caught flesh-eating piranhas in Ecuador using this method. See pages 162–6 for more on bows and arrows.

FISH POISON

Extreme stuff – and only to be used if you really know what you're doing. In some – warmer – parts of the world, there grow plants that contain a compound called rotenone. Rotenone is used commercially as a pesticide and is mildly poisonous to humans but extremely poisonous to fish and insects. Indigenous tribes have used it as a method of catching fish since the year dot.

If you know how to extract rotenone from these plants, you can sprinkle it into warm, still waters (the water temperature will need to be over 10°C) and let it work its magic. Rotenone messes with the fishes' breathing mechanisms, forcing them to rise to the surface in an attempt to get some air. When that happens, you just need to scoop them up. And because rotenone is only mildly toxic to humans, and will be present in the fish in only tiny quantities, the fish should be safe for human consumption.

Some plants that contain rotenone include:

Derris elliptica – a tropical shrub and vine. Dig up the roots and grind them into a powder. Mix large amounts of the powder with water, then throw this mixture into the river.

Anamirta cocculus – a vine from southern Asia that is also found on some of the South Pacific islands. The rotenone is contained in the seeds, which you can crush before throwing into the river.

Duboisia – a shrub native to Australia, with berries and white flowers. Crush the whole plant and add to the river.

ICE FISHING

Food can be scarce in very cold regions such as the Arctic or up a high mountain. Fish are often your best bet where there's water, but where a sea or lake or river is frozen, you need to approach fishing a little differently.

Ice can be both tough and thick. If that's the case, you'll need a tool called an auger, which is like a large metal corkscrew. If you don't have one of these, then you'll need to limit yourself to thinner ice – a few centimetres – and make a hole in it as best you can. Sharp rather than blunt objects are much better, as striking with a blunt tool will just weaken the surrounding ice. And be careful – you risk splintering the ice all around you, which will make it weaker and dramatically less safe. You must always be very cautious walking on ice – do it **only** if you're absolutely sure it will take your weight. If in doubt, don't do it. No fish is worth falling through ice for.

But if you have no choice, then here is how. Make the initial hole using the end of a sturdy wooden branch. Now make yourself a night line from a long piece of cordage, with several baited hooks along its length. Tie a rock to the bottom end of the line and tie the top end to a pole that is longer than the diameter of your hole. Lower the weighted end of the line into the hole and allow the pole to overhang the hole, as shown.

Your line is now set, but you've got a problem: if it's cold (which it is), the hole will quickly freeze up again. To slow this process down, you can shove some leafy branches (think spruce trees), leafy end down, into the hole. Counter-intuitive as it may sound, you can also then cover the whole lot over with snow, as this will help insulate the hole. Stick a little twig in the snow so you can remember where your hole is.

> Don't just rely on one of these ice holes. Set up as many of them as you can to maximize your chances of catching something. In a survival situation, you should never rely on just one food source if you can possibly help it.

PREPARING YOUR FISH

First things first: all freshwater fish are edible, but you should always cook them. (I have eaten freshwater fish raw, but you'll want to be sure that they come from very clean water before you do this, and only then in an extreme survival situation.) Freshwater fish contain all sorts of bacteria, viruses and parasites – just like the water that they live in. So just as you wouldn't drink water without purifying it first, don't do the same with the fish that live in it unless you've no other option. And if the water itself looks overly polluted, you might want to think twice about eating anything that comes out of it, cooked or not.

Saltwater fish are a slightly different matter. Generally they don't contain the same number of parasites as freshwater fish, because of the high concentration of salt in sea water. And those micro-organisms to which they do play host *need* a high level of salt, so they can't survive in your body. That's why sushi is mostly made from saltwater fish. That said, unless you really know what you're doing, it's a good idea to cook all saltwater fish too, especially if you've caught them close to the shore where they might have come in contact with fresh water.

The innards of most fish – especially the liver and the roe – are normally edible. However, there are two excellent reasons why it's a good idea to gut your fish before you eat it. In some cases, harmful toxins can accumulate in the guts and can cause very nasty symptoms if ingested. And the guts are the first part of the fish to deteriorate once it's dead, so if you remove them you can keep your fish fresher for longer. In any case, fish can go bad very quickly, especially in hot weather. Best to get it eaten quickly.

You can keep a fish alive – and therefore fresh – by tying one end of a length of string to a tree by the water, threading the other end through the fish's gills and mouth, then tying that end to a stick. The fish can then remain alive in the water without escaping, but if you're going to put them through this (and you should only do so in a genuine survival situation), make sure you intend to eat them.

KILLING AND CLEANING
Killing

No animal should be killed lightly. Only kill a fish if you're definitely going to eat it, and try to make its death as humane as possible. First off, you need to stun your fish with a blow to the head. This makes the grisly bit you have to do next a bit more humane, not to mention easier because the fish won't be flapping around quite so much. Keen fishermen will do this using a blunt wooden club called a 'priest', but a flat stone will work just as well. Don't hit the fish too hard – you want it stunned and brain-dead, but its heart still beating so that you can bleed it.

Bleeding a fish is necessary for some species whose flesh won't otherwise make good eating. But it's a good idea for any fish of a decent size because it can expel certain parasites and keep your catch edible for longer.

You can bleed a fish with your hands, or with a knife. To do it manually, hold the head in one hand, stick a couple of fingers into the gills and snap the head back. This should sever the main artery. Hold the fish by its tail to let the blood drain out: if the heart is still beating (that's why you want the fish to be alive), this should be a fairly speedy process – a few minutes or so, depending on the size of the fish.

It's better to bleed bigger fish with a knife to avoid hurting your fingers when you snap the spinal cord. Make a deep cut on the underside of the fish where the body meets the head, as if you're cutting its throat. This should take you up to the main artery. Don't make the cut too far back as you'll hit the heart and the fish won't bleed so effectively.

Cleaning

If the fish are less than about 5cm in length, you don't need to gut them. Just cook and eat them whole. Anything bigger and you'll want to gut them as quickly as possible. The process of gutting is different for round fish (think mackerel or trout) and flat fish (think plaice).

For round fish, first turn the fish upside down and locate the anal vent – a small opening near the tail end. Insert your knife and slice up to the throat. You can now scoop out the insides.

For flat fish, stick the point of your knife just below the pectoral fin. Cut sideways for about 5cm, following the curved shape of the fish's skeleton. Squeeze your finger inside the hole to remove the guts and stomach contents.

Your next job is to scale the fish. You don't *have* to do this. If you're going to cook the fish over a direct flame it's often best to keep the scales on, as they protect the flesh from burning. And if you intend to skin a fish, you should also leave the scales on as the skin comes off more easily that way.

However, if you do want to scale your fish, you'll see that the scales overlap like roof tiles. You can scrape them off using the sharp edge of a knife by running the blade from tail to head. You can also make an excellent de-scaling device by fitting two metal bottle caps to a piece of wood, the crinkly side facing out. Rub this in the same direction over the fish's skin and the scales will fall off.

> Once you've gutted a fish, don't throw away the innards. A good handful of fish guts makes ideal bait for more fishing. If you're not going to use them, throw the guts and the scales back into the water as food for scavenger fish. It's always good, if you've taken something away from nature, also then to give something back . . .

Filleting

You may not always want to fillet your fish, as it cooks very well on the bone. However, some of the cooking techniques that follow are much easier if you've filleted your fish first. And if you've caught more than you can reasonably eat in one sitting (lucky you!), you may want to salt and/or smoke the fish to preserve it (see pages 195–6). In that case, filleting's the way to go.

And just as there's more than one way to skin a cat, there's more than one way to fillet a fish. The instructions I'm about to give you should see you right for most small fish that you're likely to catch. They won't win you any Michelin stars, but nobody cares about that in the field. Filleting is one of those skills that gets easier the more you do it, so don't worry if your fillets aren't fishmonger-perfect first time round. As long as you've removed most of the flesh, you're doing pretty well.

The best kind of knife for filleting is long and slightly flexible. But the most important thing is that it should be sharp, and if you only have one knife with you, that's what you'll have to use (see pages 159–62 for more on knives). Lay your fish on its side and make an incision behind the head, cutting at a slight angle just behind the gills until your knife hits the spinal cord. Now hold the fish by its head. With the blade of the knife facing away from you and at right angles to the spine, cut along the fish's back, using the spine to guide the tip of your knife. Rotate the fish 180 degrees and perform the same action on the other side of the fillet, cutting through the belly flesh. You should now be able to separate the whole fillet from the fish by cutting it away from the spine, as shown.

SOME SMART WAYS TO COOK YOUR FISH

The trouble with cooking fish in the wild is that the direct heat of a campfire can easily scorch the flesh. Not a problem if you happen to have frying pans and grills with you, but if not you have to think of ways around it. Fortunately there are loads of cool ones.

PONASSING

This is an extremely old way of cooking a whole fish. It's great for campfire cooking, uses equipment that you can easily find in the field, and once you've mastered a few basic techniques it's very easy. Best of all you get a delicious piece of fish with a great smoky flavour – it'll do justice to all the hard work you've put into catching your supper.

Preparing your fish

To start with, you'll need to butterfly your fish. This is a way of filleting it so that the two fillets remain joined together.

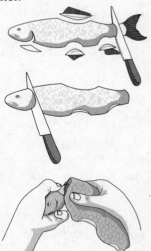

First, gut and clean your fish in the usual way. Cut off the tail and remove all the fins.

Now, cut around the fish's neck, down to – but not through – the spine.

Hold the fish in one hand. Press the thumb of the other hand into the cut and against the spine. Now gently pull the head and spine away from the fillet. You should find that the spine and ribs come easily away from the flesh and you'll be left with a butterflied fish.

Fixing your fish to a frame

You'll need a sturdy, straight piece of wood about 60cm in length. Hazel is good – it's non-toxic and won't burn easily. Sharpen one end so you can stake it into the ground. Use a small axe or saw – or at a push, your knife – to make a split along the other end. It needs to be the same length as the fish.

Now you'll also need to fashion two wooden skewers from some thin, straight twigs, each one about the same length. Remove the bark and sharpen the ends if you like. Make four holes in the fish and skewer it as shown.

Slide the fish into the split stake, then tie the split end to hold it in place.

Make a fire. When you have hot embers, drive the stake into the ground at an angle so that the fillet is suspended above the fire. How long you need to cook it for depends on the size of your fish and how hot the fire is, but 10 minutes normally does the trick.

WET NEWSPAPER

This is a really easy one. Take a whole gutted fish and wrap it in five or six sheets of wet newspaper. Place the whole thing in the embers of a fire. When the newspaper's dry, your fish should be cooked – you'll know if it is because the flesh will flake away easily from the bone. Unwrap it and dig in.

An alternative to wet newspaper is to wrap the fish in sturdy, non-toxic leaves. Vine leaves or banana leaves are good. Use several layers to make a tight parcel and tie with string or natural cordage before nestling it in the embers of a fire, carefully placing some more hot coals on top and cooking for 10–15 minutes, depending on how big your fish is.

PLANK COOKING

You'll need some decent axe and knife skills for this. Have a look at my book *Living Wild* for the low-down.

Find yourself a log of non-toxic wood – oak would be great. Carefully use your axe to split the log in half lengthways.

The plan is to peg a fillet of fish to the cut side of one of these halves. Make some pegs by sharpening a few splinters of wood, then use your knife to create some 'starter' holes in the plank. Lay your fillet over the holes, then tap the pegs through the fish and into them.

Light yourself a fire. When you have burning embers, lie the plank close to it, but not too close. If you can comfortably hold your hand in the vicinity for a few seconds, you've got it about right. Prop the spare half of the log behind your plank so that the fish faces the fire. The fish should turn deliciously brown in a few minutes.

Finding fresh water by the coast can be a challenge. But if you've managed to catch a saltwater fish, you can extract water from its flesh. Cut the fish into small pieces, then put them in a piece of clean cloth and wring the moisture out. The liquid that you get is not too salty and can keep you alive when water is scarce.

HOT ROCK COOKING

Find yourself some smooth, flat stones. Lay them on the ground to make a solid bed. Light a fire on top of the stones and let it burn down to hot embers. Now – carefully – find something to brush away the embers from the top of the stones. You could use a leafy branch to do this, but not a very dry one that will ignite. You can now lay your gutted fish or fish fillets directly on to the hot rocks, skin side down. You'll find that it cooks through very quickly.

Fish Head Soup

If you've caught a fish, don't let any of it go to waste. There's plenty of good nutrition in the head. Simply boil it up in a little water. You can then pick off any scraps of meat from the cooked head – don't forget the eyeballs, which are edible – then drink the broth, which will be infused with fish oils, protein and vitamins.

DANGEROUS . . . BUT (MOSTLY) EDIBLE!

CATFISH

As we've already said, you can eat any freshwater fish as long as they're cooked. You need to be careful, though, of the catfish. These are found in the inland waters of all continents except Antarctica and are particularly abundant in tropical regions. They especially like muddy rivers, and there are also some saltwater varieties. They are good big fish with plenty of meat on them, and they are not normally aggressive, but many species have stings near their fins that can cause extreme pain and inflammation. If you catch a catfish, handle it very carefully.

PUFFERFISH

Pufferfish is one of the most poisonous vertebrates in the world. You sometimes find them in temperate waters, but they are much more common in tropical zones. Their liver, ovaries, intestines and skin contain a neurotoxin that can be fatal to humans. That doesn't stop some people wanting to eat them, but it's a rare skill to prepare a pufferfish in such a way that consuming it won't be fatal. I *don't* recommend you try it in the wild!

STINGRAY

The stingray is a lethal reef predator with powerful jaws and a venomous sting in its tail. They are found in tropical and subtropical waters, though they might sometimes put in an appearance in warmer temperate waters. They're edible, but you need to be careful of their venomous barbed tail, which has been known to kill.

5

WILD FOOD
BY THE SEA

The indigenous people of Alaska have a saying: when the tide is out, the table is set. It's true. You might look at a wide expanse of beach after the tide has receded and think that it doesn't offer much in the way of nourishment. But really you just need to know how and where to look. Wherever you are in the world, from the frozen wastes of the Arctic to the burning shores of Australasia, the seashore can be a life-giving source of ready nourishment.

This chapter is about foraging at the water's edge, and all the good, nutritious food to be found there.

STAYING SAFE BY THE SEA

If you're anything like me, you can spend hours by the seashore. And that's fine. Just be careful. The tide might look a long way out, but it can come in very fast indeed. And tides are local – just because you know what the tide is likely to be like in one location, it doesn't follow that you'll know how it acts a few miles down the road. Arm yourself with local knowledge and always keep one eye on the water. And remember, there are few forces in nature as strong as the tide. Don't get caught out.

If you find yourself foraging around large rocks by the seashore and the waves are fierce, watch them for a good half hour before venturing into the water. This should be sufficient time to tell you which areas of the rocks are likely to get pounded by the waves. Avoid those areas, because you'll get pounded too.

SEAWEED

Seaweed might *look* a bit disgusting, but in fact it's one of the best trail foods out there. Generations of people have used it all across the world as a survival food. It's easy and quick to gather and mostly safe to eat. Perhaps more importantly, you can preserve it to provide long-term nutrition. Some of it also happens to taste really great. What's not to like?

The only poisonous seaweed is the *Desmarestia* species, but this probably won't be a problem since it generally only occurs in very deep waters – too deep for foraging – so you're unlikely to come across it.

Otherwise, seaweeds are a fantastic source of minerals and protein, and particularly of iodine – an essential mineral that occurs naturally in hardly any other foodstuff. You could eat all seaweeds except *Desmarestia* in a survival situation, but many of them are prized as gourmet treats by different cultures around the world and are even cultivated for the pot. The good news is, you can just pick it wild. You simply need to follow a few guidelines:

1. Harvest your seaweed fresh from the water. Don't eat any dodgy, smelly stuff that's been lying around on the beach for a while and avoid seaweed that comes from polluted waters near sewage pipes, harbours and ports.

2. Carefully pick your seaweed over for small stones and shellfish. Then wash it thoroughly in fresh, purified water (you'll be amazed how much sand comes off it).

3. Some seaweeds are palatable raw, but if you're able to, you'll probably want to boil it before eating it.

In the pages that follow, I'm going to tell you which of the seaweeds that you're likely to find make good eating. I'm also going to show you how to dry your seaweed in the wild, and how to turn it into an amazing, portable survival food.

If you can, harvest your seaweed by cutting no more than a third from the top of the plant. That way, the plant has a chance to grow back.

If you think you are likely to be harvesting some seaweed, take a string bag with you – otherwise you'll find yourself carrying around a lot of (heavy!) sea water.

Seaweed is fantastically good for you, but beware: if you eat a lot of it on an empty stomach, you might suffer from diarrhoea – not fun in the wild . . .

SEA LETTUCE
(*Ulva lactuca*)
Where to find it: worldwide.

You've probably seen sea lettuce any number of times. It looks like – you guessed it – lettuce floating in the sea. Sea lettuce is a very intense, almost luminous green. It's very high in protein, fibre and iron. You can eat it raw as a kind of 'seaweed salad' (as long as you don't mind chewing for a while), fry it, cut it up into soups or stews, toast it in a dry pan over your campfire, and air-dry it (see page 114).

DULSE
(*Palmaria palmata*)
Where to find it: Atlantic and Pacific coastlines.

This is another seaweed that you've probably seen a hundred times. It has a deep red colour and spreads out in broad blades, like a tangle of wide, flat, red pasta. Dulse has a very high protein content and contains all the trace elements needed for human survival. A real super-food. Like the sea lettuce, you can eat it raw, but you'll have to chew it even harder and longer to make it digestible. In Iceland they eat it with butter, but you can also chop it up and add it to bread dough or campfire stews to add flavour. It's very commonly dried, when it will become very crisp and easy to eat – a great, nutritious snack for when you're on the move.

GUTWEED
(*Ulva intestinalis*)
Where to find it: worldwide.

Horrible name, but an awesome wild food simply because it's so common (although in Japan they also cultivate it). It's an amazing bright green colour with long, tubular fronds, and on sunny days the gutweed fills with oxygen produced during photosynthesis. You can eat it raw – it makes a great salad dressed with a bit of oil and vinegar if you have them in your pack. It's also excellent dried and crushed: you can use the resulting flakes to flavour your campfire food (they are particularly good sprinkled on top of a wild pizza – see page 80).

CARRAGHEEN
(*Chondrus crispus*)

Where to find it: mainly in the North Atlantic, but has been recorded worldwide.

Also known as Irish Moss. Believe it or not, you've probably eaten this before, because carragheen extracts have been used on a massive commercial scale to thicken ice cream. You can use it for the same purpose by drying it then grinding it to a powder, before reconstituting it with water – like gelatine. It stands out from other green seaweeds because of its very distinctive colour, which ranges from red to dark purple or brown. Out in the field, you can also use it to thicken a soup or a stew simply by adding a little dried carragheen to the pot and simmering it for about 25 minutes. It doesn't have a particularly strong taste, so don't worry about it affecting the flavour of your cooking.

KELP
(*Laminaria digitata*)

Where to find it: worldwide.

Another super-healthy seaweed, kelp is also super-sustainable – it can grow up to half a metre a day, so if you harvest a little for your breakfast, chances are it will have regenerated for your dinner. It is dark brown in colour and has a smooth structure that is difficult to break. Like the other seaweeds, it dehydrates really well, but it comes into its own if you take it home and turn it into seaweed crisps. Make sure your kelp is nice and dry, then cut it into 4cm squares and deep fry for a few seconds in oil that has reached 180°C. Drain on kitchen paper and eat quickly.

LAVER
(*Porphyra umbilicalis*)

Where to find it: Atlantic and Pacific coastlines.

If you've ever spent much time in Wales, you might have heard of laverbread, which is cooked and puréed laver. It's also the seaweed that the Japanese use to make nori. Laver itself doesn't look that appetizing – almost like a gooey membrane covering large seashore rocks. But there's nothing to stop you making laverbread in the wild.

Laverbread

Rinse your seaweed well, put it in a pot and cover it with cold, fresh water. Bring to the boil, then simmer for a couple of hours, stirring it regularly and making sure it doesn't dry out. Add more water if necessary. When the laver is very soft, strain off any remaining liquid and mash it to a pulp using a spoon or a clean stick. If you've brought oatmeal with you for your morning porridge, try making a mixture of one third oatmeal and two thirds laverbread. Roll it into little balls and fry in some oil for a fantastic breakfast.

How to dry your seaweed

As you've read, lots of these seaweeds benefit from being dried. It makes them super-portable in the field and means you can have both a nutritious snack to hand and a fantastic wild flavouring to add to your cooking. But you do need to dry it carefully. Here's how.

1. Rinse your seaweed in a freshwater source. This gets rid of salt, sand and any shells or stones that have stuck themselves to the seaweed. See page 22 for the low-down on fresh water, but bear in mind that you can do your first few rinses directly in a stream, before doing the final one with purified water.

2. Make yourself a little washing line. Tie one end of some cordage to a tree, the other end to a stake in the ground, and hang your seaweed on it.

Alternatively, erect a straight branch between two trees like this:

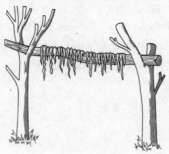

3. Dappled shade and a light breeze on a sunny day are the best conditions in which to dry your seaweed. It should be done in a couple of hours, but make sure it's fully dried within 12 hours otherwise any remaining moisture will make it start to go mildewy. Once it's dried, store it somewhere dry and it should keep for a couple of years.

Seaweed Soup

Seaweed makes a good addition to any vegetable or wild-green soup you might be cooking upon your campfire. Since the leaves don't break down like other greens, you probably want to chop it up before adding it to the pot. And you can make a soup out of just seaweed itself. The easiest way to do it is to take about half a litre of fresh, purified water. Bring it to the boil and add a stock cube. Then throw in three handfuls of whatever fresh, cleaned seaweed you have to hand. How long you cook your seaweed depends on what type you use and how thick and rubbery it is. Laver will take a good 90 minutes, whereas sea lettuce will be done in just a few minutes. You're aiming to get it nice and soft, otherwise you'll be doing a lot of chewing.

If you have some dried dulse with you, rehydrate it in some fresh water, then boil it for 10 minutes and chop it up very finely before adding it to your soup. It will give you a really rich taste.

Wild Sushi Rolls

I can't promise that this is going to look as perfect as the stuff you find in sushi bars, but if you can get your hands on some sea lettuce, you can wrap it round almost any other food to make a really substantial and delicious wild snack. Simply pile chunks of cooked meat or fish in a leaf of rinsed sea lettuce and wrap it up. Chewy but tasty!

MOLLUSCS, CRUSTACEANS AND OTHER SEASHORE GOODIES

The main edible molluscs that you're likely to encounter are the bivalves: mussels, cockles, oysters and clams. If you've ever come across a mussel bed at low tide, you'll know how incredibly plentiful they can be – a real life-saver in a survival situation.

You can, however, make yourself *very* ill from eating these molluscs. Bivalves filter gallons of sea water every day and can harbour all manner of bacteria, viruses and poisonous algae. We're talking *E. coli*, Norovirus (both of which will reacquaint you with your supper much sooner than you might expect) and the potentially fatal paralytic shellfish poisoning. So before you start collecting them, you need to know how to minimize your risk.

The best way to establish if it's safe to pick molluscs from a particular beach is to ask a local. Of course, if you're miles from civilization that might not be possible, so here's what you need to do.

1. Only harvest molluscs from clean waters. Avoid areas near harbours or ports and, especially, sewage outlets. You don't need me to tell you why.

2. Pay attention to the old northern hemisphere saying that you should only harvest molluscs if there's an R in the month (i.e. avoid the months of May to August – summertime). There's a good reason for this. Poisonous algae flourish when the water gets warmer. Stick to foraging molluscs in the winter and you should be fine (this is especially true in tropical areas) – unless, of course, you're in a genuine survival situation and there's no other food source available.

3. If a mussel, oyster or clam is open when you pick it, give it a tap. If it doesn't close up, it's probably already dead and you shouldn't

eat it. Likewise, when you've cooked your shellfish, discard any that haven't opened up, for the same reason.

4. *Always* cook molluscs. Cooking kills pretty much all bacteria and viruses, which means you only have to worry about poisonous algae. In a survival situation, forget about steaming lightly to make a creamy *moules marinière*. Boil or steam your shellfish for a good 5 minutes before eating.

MUSSELS

The most common, and probably the tastiest, of the molluscs. They're native to all oceans, but thrive best in temperate waters. Unfortunately, they're probably the worst carriers of shellfish poisoning. Make sure you follow the guidelines above before eating them.

You'll find mussel beds in what's known as the intertidal zone of the beach – the area between high tide and low tide – so make sure you keep an eye on the tide when you're gathering them. You can't miss them – vast sheets of shellfish spreading out sometimes for hundreds of metres. The big ones are the best, and are normally found nearer the low-tide mark.

To prepare mussels, rinse them first in fresh water. You can, if you like, soak them for 48 hours in clean salt water to make them less gritty, but this can be hard to achieve in the wild. Discard any that don't close when tapped, and pull off their tiny 'beards' (tough little strands attached to the shell). Don't worry too much about scraping off the barnacles, though. Put the mussels in a pot with a little boiling water, cover and steam or boil over a hot flame for at least 5 minutes. When they're cooked, discard any that aren't opened.

COCKLES

Cockles are found all over the world. They are much smaller than mussels, but still a good source of wild food because they occur in vast quantities. Like mussels, these are found in the intertidal zone of the beach, normally semi-buried in the sand. You might find that you need a small rake – or a clawed hand – to uncover them.

Cockles can be full of sand. The best way to get rid of it is to let them soak in a container of clean sea water for a few hours, then cook them in the same way as you cook mussels.

CLAMS

There are many different species of clam, but they're all edible. Like cockles, they hide under the sand in the intertidal zone. You can tell they're there, though, by little dimples (called siphon holes) in the sand. Uncover them by using a rake (or your hands) and cook in the same way as mussels.

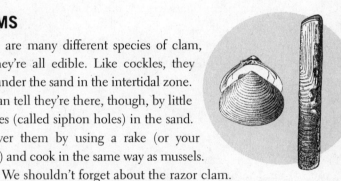

We shouldn't forget about the razor clam. You might have seen empty razor clam shells along the beach, long and pearl-like. To get them when they're still full of meat you have to be a bit cunning. The presence of razor clams is indicated by a hole in the sand shaped like a keyhole. If you pour a little ordinary table salt (perhaps you have some in your trail spice box) into one of these holes, the razor clam will emerge suddenly from the sand. Grab it quickly, before it disappears again. Pull gently and the whole clam will come free of the sand, with the meat hanging below the shell. Rinse the clams in clean water, then boil for 5 minutes. Best to avoid the black stomach – a small, distinctive sac attached to the main body of the clam – which can make you ill, but otherwise razor clams are delicious.

LIMPETS

You'll find these clinging tightly to rocks at low tide. They're very hard to remove – you'll certainly need a knife to wedge between the shell and the rock so you can prise them off. (If they're *not* hard to remove, that's a bad sign and you should avoid them – though you'll probably find that the tide has washed away any dead or ill limpets.)

Once you've got a decent harvest, boil them for at least 5 minutes, then cut them away from their shells. You'll find you have the internal organs and a lump of orange flesh – discard the organs, eat the flesh. That said, the method I use most is just to knock them off the rock with the pommel of my knife then roast them in the embers of my fire. Delicious!

OYSTERS

If you've found an oyster bed, you've really lucked out. They not only taste good, they're also large and nutritious – an amazing survival-food find – and are found in all oceans. Just don't be tempted to eat them raw, like you might in a restaurant, unless you have no other option. Cook them well, just as you would mussels. (This has the advantage that you don't have to 'shuck' them open when they're still alive – very difficult without the right tool. Steam or boil them for 5 minutes and you should find they've opened by themselves.)

All the molluscs above make great eating, but they're also very good as fishing bait – see pages 89–90 for the low-down.

SEA URCHINS

If you tread on a sea urchin, you'll know about it. They're like spherical porcupines, with hundreds of sharp spines – very hard to get out once they've embedded themselves in your skin. There are lots of different kinds – some Mediterranean sea urchins are pitch-black, whereas in the UK the most common edible one is brick orange in colour. You can find them among seaweed and rocks in all oceans, where they feed on algae and small shellfish.

To catch a sea urchin, pierce it with a sharp stick, then hack away the sharp tips of its spines with your knife. The only edible parts are the gonads and the eggs. You get at them by splitting the urchin in half with your knife and scraping away the sand and the guts. The gonads are soft, squishy brown sacs. The idea of eating a sea urchin's gonads might not be that appetizing, but they're very high in energy. In Japan they rinse them in water and eat them raw. You could try that, but if you can it's probably safer to boil or steam them before eating. (Some people call sea-urchin gonads 'sea caviar'. I can't promise that you wouldn't prefer *real* caviar, but they're a great survival food in any case.)

BROWN CRABS

Of all the crabs, the brown crab will provide you with the most meat when it's of a proper size. Unfortunately, most of the big ones tend to hang out in very deep water, but you'll often find smaller specimens in shallow waters or in rock pools at low tide and they make a great food source.

Large brown crabs do sometimes hide in little rock caves, however, waiting for the next tide to come in. You can coax them out by using a hook on a stick (one of the improvised fishing hooks on pages 85–7 would do the trick). If you're lucky, a hiding crab will catch hold of the hook. Very gently pull the crab towards the mouth of the cave. When you can, grab the top of its shell with your free hand.

Cook your brown crab in boiling sea water (this is humane, as it will kill it instantly) for between 15 and 25 minutes, depending how big it is. When it's cooked, let it cool then twist off its claws and legs. Turn it upside down and pull away the body from the shell. Remove the 'dead men's fingers' (grey spongy bits), then scoop out the brown meat from inside the shell. Cut open the body lengthways and remove the edible white meat from inside it. Now crack open the claws and legs. Remove and eat the white meat they contain.

SEA CUCUMBERS

Despite their name, sea cucumbers are creatures not plants. They live on the sea bed in all oceans, but you can sometimes find them washed up under rocks or in the sand. They look like black cucumbers covered in little nodules, and they grow up to about 20cm. A great protein source – I've eaten them raw, which is pretty gross, but if you can, you should boil them for 5 minutes before splitting them down the middle and eating the innards. If they don't move when you touch them, or if they smell bad once cooked and split open, avoid.

OCTOPUS

Hard to catch, but these molluscs are found in all oceans and sometimes venture up to shore at night, especially in warmer parts of the world, when you can lure them with a torch. If you find one, spear it with a sharp stick then get it to shore as fast as possible.

You can kill an octopus quickly by driving a knife between the eyes. You can also efficiently kill one by turning its head inside out and pulling away the innards. (This takes a bit of practice – try it on a small one first!)

Octopus flesh is very nutritious and you can eat it raw – but be warned: I once did this and the suckers on its tentacles were still working, which meant they stuck to my lips and my throat as I swallowed. Best to boil the body till it's cooked through, though the tentacles are also pretty fine roasted over an open fire.

If you're foraging around the Pacific and Indian Oceans, especially round the Australian and Tasmanian coasts, look out for the blue-ringed octopus. It's not large – it only grows up to about 20cm in length – but it is very distinctive. When resting, it has a beige colour with brown patches. But when it gets agitated, those patches turn a beautifully bright, vibrant blue. When that happens, get out of there. It's very venomous, will bite if provoked and can easily kill a human. It has a bite that can puncture a wetsuit without any difficulty and its venom causes total motor paralysis. There's no known antidote.

SEA BIRDS

Birds can congregate in vast numbers by the seashore. Obviously they'll fly away if you get too close, but in a survival situation it's possible to catch them much as you'd catch a fish – and it's worth remembering that all birds are edible. Attach a bit of bait (offal is good) to a hook tied to a long piece of fishing twine and lay it on a rock. Wait patiently: chances are a bird will try to eat it. When it does, give a good hard tug to catch the birdie. See page 190 to learn how to despatch your bird once you've caught it.

Alternatively, if there are birds overhead you can throw a baited hook up in the air for one to catch while it is flying. (You might want to tie a small stone to the line so you can get a bit of height.)

You'll also find some other good ways of catching, killing and preparing birds on pages 154–7 and 189–91.

If thirst is a problem by the sea in a genuine survival situation (and trust me, you don't want to drink sea water, no matter how tempting it might be), then you can drink the blood of a bird or turtle to stave off dehydration. (If you read my book *True Grit* you'll find the amazing story of Louis Zamperini, who stopped himself dying of thirst by drinking albatross blood when he was stranded at sea for nearly fifty days . . .)

A REALLY COOL WAY TO COOK A WILD FEAST BY THE BEACH

If you've managed to forage a good haul of seaweed and shellfish, and if you're pretty certain that it's all of good quality, you could try a clam bake. These are very popular in New England, where they're reserved for special occasions. When you're hungry in the wild, a feast like this will really boost your morale.

As well as your seafood and seaweed, you'll need a piece of canvas tarp that you've soaked well in sea water, some round stones and some firewood. First off, dig a fire pit on the beach (you may need to check that you're allowed to do this first) and layer the bottom with the stones. Light a fire on top of the stones and let it burn for a couple of hours until you have a layer of good, hot embers over the stones. Carefully brush away the embers to the side of the pit, then cover the stones with a good layer of seaweed, which will act as a protective layer and produce lots of steam. Add a layer of seafood (plus any other food that you want to cook – potato chunks, vegetables, bits of meat), then more seaweed, and continue layering until you reach the top of the fire pit, finishing with a layer of seaweed. Now place your wet tarp over the top and leave for at least 2 hours. The food will steam deliciously in the pit and when it's cooked you can, literally, dig in.

PART TWO

GETTING YOUR HANDS DIRTY

6

STALKING AND TRACKING GAME

On the lookout for wild food in the swamps of the Northern Territories of Australia. (In this place you're not top of the food chain.)

Above: Improvising hunting weapons is a key skill in the wild. This spear would be ideal for hunting small mammals. The stick in my left hand triples the velocity of the spear and turns it into a deadly weapon.

Below: On a raft I made in the jungle of Belize — a perfect place to trail fishing lines from.

Real salmon sushi,
Alaska-style.

Above: Edible frogs, the salvation of many a survivor. (Beware of brightly coloured frogs, and all toads.)

Below: What goes through my mind before eating critters like this is generally: Oh s***!

Left: Blend into your environment, stay alert and use your superior brain in the hunt for food.

Above: Snakes are always risky to catch, but a great source of protein and energy.

Below: A large boa constrictor that I caught in Belize. It fed me and all the crew for three days!

Inset: This alligator was six foot, and was just about manageable to catch and despatch.

The skull of a large saltwater crocodile.
The power these guys have is off the scale.
Do not get in a fight with one of them.

Raw zebra meat, on the bone — it doesn't get better than this (although this lion kill was fresh and I didn't want to hang around long to fight them for it).

The natural world is like a book: full of information for those who know how to read it. Learning how to read the signs of nature is just like learning to read words. You start off with your ABC before moving on to more complicated texts. And if you keep reading and keep learning, over time you'll start to see what few people ever get to observe.

In the hunt for wild food, one of the most important chapters in the book of nature is the one that tells you how to track wild animals. I'm not going to dress this up for you. Knowing how to stalk and track game can be a harmless and fun pursuit, but in extreme survival situations it is the prelude to something that can't be taken lightly: the death of an animal.

I don't want you to be gung-ho about this. If you can survive by foraging wild plants or even insects, try to do that before you think about killing animals. It's not only that you'll expend a lot more energy hunting game. You should only take an animal's life in the wild if you have no other option. I don't eat a great deal of meat at home, and I try to minimize it in the wild too.

Having said that, the person who can stalk and track wild animals will often come off better in the fight for survival than the person who can't. The flesh of all furred mammals is edible, and the same goes for the flesh of all birds. (Although just because it's edible, it doesn't necessarily follow that it's tasty.) So the better you can learn how to 'read' the wild *before* you find yourself in a survival situation, the greater your chances of survival will be should the time arrive when you have no option but to kill your quarry for food as humanely as possible.

What follows is the ABC of tracking. We'll concentrate mostly on furred mammals because these are the creatures you'll most likely be tracking for food in a survival situation. But once you've mastered this, I hope you'll go on to learn a lot more: trust me, this is one of those fields of study where you never stop learning. Perhaps one day you'll become like one of those awesome aboriginal scouts who can track animals (and humans!) over vast distances, using signs on the landscape that only they can see . . .

ANIMAL TRACKS

In the chapters that follow, you're going to learn a lot about how to trap, kill and cook wild animals. But before you do any of these things, you have to find them first. Knowing how to stalk and track your prey by recognizing how to follow their tracks is one of the most important survival skills you can learn – if you want dinner, you'll have to find it first.

Tracking takes real skill. Don't imagine that you're going to find yourself in a survival situation and suddenly start seeing animal tracks all over the place. You need to practise examining the terrain very closely, and often you'll need to learn how to piece together the evidence of partial prints. In fact, unless you're tracking an animal through the snow, wet ground or damp sand, it's pretty unlikely you'll ever come across full prints. Have fun piecing all this together when you're out on the trail and you'll be far more likely to have the skills to hand when you really need them.

SOME THINGS TO CONSIDER WHEN YOU'RE TRACKING
Landscape

Think about what an animal needs to survive: food, shelter and water. The most likely places to track wildlife are those where there is plenty of mixed vegetation and enough undergrowth for animals to hide and shelter in. Wide open spaces or deep forest aren't so good. Your best bet is to head to what are known as 'transition' areas – where one type of terrain meets another, such as a field and a forest, or a forest and a stream. These areas are host to a mixture of vegetation and cover, and are therefore attractive to a wide range of animals.

And while it's true that animals will often beat a path to a water source, a stream, river or lake is not always necessary: many herbivores get the water they need from early-morning dew or from the vegetation that they eat.

Paths

When an animal passes through high vegetation, it will leave a path. Examine this carefully: the vegetation will be bent in the direction the animal travelled. Sometimes, of course, these are established paths and don't necessarily indicate that an animal has *recently* passed by. But follow them anyway – there's a good chance that they'll lead to a den or a regular feeding or watering area, which are very good places to stake out in your hunt for wild food.

Broken twigs on the ground ahead are a good sign that an animal has passed that way. Examine the break in the stick: if the wood is still green and fresh, that's a strong indication that the animal has passed by recently.

Poo and pee!

Once you get your eye in, you'll start seeing animal faeces all over the place. If you can learn to recognize the very distinct types of droppings (or 'scat', as it is often called) that different animals produce, you're well on your way to finding the animal itself. You could spend a lifetime learning about animal scat, but the main thing to remember is that the presence of poo means an animal has been this way. The fresher the poo, the more recently the animal has passed by.

Herbivores have to eat a lot more food than carnivores to get the same amount of energy. This means they produce more faeces. Herbivore droppings (think rabbits, hares and deer) are generally small and round; carnivore droppings (think dogs, bears, foxes) more tube-shaped.

Be careful if you're examining animal poo. It can harbour germs, so don't touch it. Even when it's dried out, it can give off a dust that can carry and spread disease. Stand upwind.

Animal pee is a slightly different matter. There are expert trackers in the world who claim to be able to smell it on the wind. That's a pretty advanced skill set! I prefer to stick to hunting for poo.

Dens

The best place to find an animal is at its home. If you've spent any time in the wild – or simply been on a few country walks – you'll probably have spotted holes in the ground that are clearly animal dens or warrens. However, you can't just sit outside an animal hole and expect dinner to turn up automatically. Animals move around a lot, so many of the holes that you'll find will have been long abandoned.

Look for the signs – most of them are common sense. If there's a spider's web over the hole, for example, or vegetation growing over it, chances are nothing has come in and out of that hole for a while. On the other hand, if there are fresh faeces nearby, or signs of feeding (teeth marks on vegetation, for example), then you can be fairly sure there are animals – and therefore food – about.

You might be hunting animals that don't live in dens. But they still have to rest, and often leave marks on the landscape that show where they've done this. Deer and hares, for example, sleep on beds of thick vegetation, often in well-camouflaged areas. You can sometimes find the imprint their bodies have made – a sure sign that the animal itself is probably very close.

Think like an animal

If you think that your quarry might be in the vicinity, be aware that wild animals are very sensitive to sound, smell and movement. A rabbit in its warren will never emerge if it can hear your heavy human footsteps on the ground above. Walk quietly, and not with flat feet – putting your weight gently down on your heel first is a quieter way to walk and also stops you stumbling. When it comes to moving stealthily, I like to say: sight it, place it, weight it. Move slowly, stop frequently. Listen hard. When you're hunting, your ears are often more important than your eyes.

Dawn is the best time to hunt, because this is when most animals are out and about. If you can't avoid hunting at night, set out before nightfall so your eyes can get used to the dark. Be aware, though, that at night most animals will be able to see you better than you'll be able to see them.

Remember that you absolutely stink to another animal, so you have to be very aware of the direction in which the wind is blowing. Move *against* it to avoid your smell being immediately carried towards your prey.

If you're waiting for an animal to appear, camouflage yourself well and remain *absolutely* still. Remember that many animals spend most of their time trying not to be eaten by other animals. Any sudden, unexpected movement will scare them away immediately. And if your quarry *does* see you, freeze. Often that will be enough to stop them being spooked into flight.

I'm now going to teach you how to recognize a few basic tracks of certain mammals that you might find yourself stalking for food. You'll find many of these animals, or their relatives, in temperate, tropical, desert and Arctic regions. Once you've learned how to start spotting these, you can start building up your own library of tracks.

If you find an animal track, examine it carefully to see how fresh it is – there's no point trying to track an animal that passed by a week ago. A very clear track is likely to be more recent than a less clear one. If it has been raining and there are no raindrops on the track itself, the animal passed by after the rain stopped. Similarly, if there's dew all around but not on the track, it passed by after the dew settled.

Animal tracks are easier to spot when you're moving uphill. And remember that an animal probably won't step straight down, but heel first – this affects the shape of the print.

DOG

No, I'm not really suggesting that you eat Fido (though plenty of people in the world do, and in a survival situation you've got to do what you've got to do. In fact, many of the great

Antarctic explorers took dogs with them as a food source), but knowing how to recognize a dog's footprint can help stop you following a false trail, because you'll often come across them in the wild.

RABBIT

Where you'll find them: every continent except Antarctica; a serious pest in many parts of the world, especially Australia.

Easy to recognize because their hind prints make a much deeper impression in the terrain than their fore prints. Rabbits are found widely all over the world, mostly live in burrows and tend to use the same trails, which gives you a good idea of where to hunt for them.

They're one of the easiest animals to trap because they're abundant and can be taken using a simple snare (see page 149), but remember this: you can starve to death by eating just rabbit. Their meat lacks certain vitamins and minerals and so your body will use its own store of these to digest the meat. You *must* eat some vegetation along with a rabbit diet.

Myxomatosis is a common disease amongst rabbits in certain parts of the world. Affected animals will have swollen glands, be slow and sometimes blind. They look off-putting, but are in fact OK to eat. Just make sure you skin them first (which you would do anyway) and avoid the liver, which might have little white spots on it.

HARE

Where you'll find them: Europe and North America; the Arctic hare is common in the tundra, and the desert hare in desert and semi-desert environments.

If there are hares in the vicinity, you might find more than their paw prints – they rest in the open air, so you might also find an impression of their whole body. Like rabbits, they are common in many parts of the world and fairly straightforward to trap in a simple or sprung snare (see pages 149–50).

SQUIRREL

Where you'll find them: tree squirrels are found on all continents except Antarctica; ground squirrels (also known as marmots, prairie dogs or chipmunks, depending on their size) are also very common across the world.

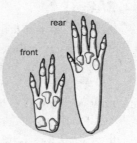

Squirrels are an abundant source of wild food. Their rear paws look a bit like a human hand with four 'fingers' and a 'thumb'. The front paws have only the four digits.

DEER

Where you'll find them: worldwide; in the Arctic, think reindeer (or caribou); other deer-like creatures include antelope and elk.

Deer prints vary a little according to their species. They all have cloven feet, which form two teardrop-shaped halves.

Deer and their relatives – antelope, gazelle, elk, reindeer and the like – are found all over the world. They are incredibly shy, and will run a mile if they see you, but don't be tricked into thinking they won't attack if they need to. And trust me: you don't want to be gouged by a deer horn.

They can be snared or caught in deadfall traps (see pages 152–3), but as they vary in size you will have to set the size of your traps accordingly. If you manage to catch a deer in a survival situation, you'll have many days of good-quality meat on your hands, and it preserves very well (see Chapter 10). Having said that, they're pretty tough to catch and you're generally better off going for small game if possible.

SKUNK

Where you'll find them: North and South America.

Skunk tracks have five toe prints on each foot, and you'll normally see deep claw marks in front of the toes. Be careful when tracking skunks. You probably already know this, but they spray a foul liquid from their anus when they're threatened, and the smell can travel for at least a mile and will stay on your clothes for months. If it does get on your clothes, the smell is brutal to have to walk around with – I know from experience! Also, if you get this skunk spray in your eyes, it will hurt and temporarily blind you. If that's not bad enough, skunks can carry rabies. If you snare one, approach it with caution, and get it killed before it can bite you. Make sure you cook it very well to destroy the rabies virus.

All in all, not the easiest dinner you'll ever catch. Oh, and the meat tastes pretty disgusting – although that might be because I still had the foul spray residue on my hands after I caught it and then cooked it!

BEAVER

Where you'll find them: Europe and North America; occasionally in the Arctic.

As aquatic creatures, beavers have webbed hind feet. You might also sometimes see a mark on the ground where their tail has dragged behind them.

They're very good to eat. The tail is especially meaty and nutritious. Look out for their dams and set up spring snares or deadfall traps nearby (see pages 150 and 152–3). You can locate their regular runs by looking for chewed trees in the vicinity.

VOLE

Where you'll find them: pretty much everywhere!

Voles are extremely common and make a good mouthful of protein if you can catch them. They are also a good 'indicator species': lots of other animals eat them, so if you see vole prints, chances are there will be other species in the vicinity.

In general, it's a good idea to get rid of any qualms you might have about eating small rodents. Mice and rats occur almost everywhere – catch them in a bottle trap (see page 154) and you've got yourself a great food source in a survival situation. Just be careful not to split their guts open, and make sure you cook the meat very well – rodents can carry diseases that are harmful to humans.

FOX

Where you'll find them: widespread across the northern hemisphere; Arctic fox in Arctic regions; sand fox in desert regions.

Fox prints are very similar to dog prints, but narrower, and the claw marks are a little longer.

Foxes and other wild dogs will smell you a mile off, so it's pointless trying to stalk them. They can be caught in deadfall traps (see pages 152–3) – but as with all animals, be very careful if you approach an injured one. A wounded fox will be aggressive;

their teeth are sharp as needles and they can bite you very badly.

Fox meat is very tough, so if you do end up eating it, you can tenderize it first by soaking it overnight in salt water.

WILD PIG

Where you'll find them: Europe and Asia.

Wild pigs come in all sorts of shapes and sizes and tend to live in forested areas. Where you find one you'll probably find more, because they live in family groups. They're easier to stalk during the day than at night, and they are often very aggressive – especially if they're guarding their young. Their tusks can inflict serious injury if they go for you, and they probably will if you wound but don't kill them. (I've been in a few battles with wild pigs before and you do not want to take them on unless you are trained and you have a back-up plan – otherwise you'll end up skewered by a tusk.) So choose the trap that is most appropriate for the task – a pitfall trap (see page 153) or heavy-duty snare is best. Then spear it from a distance.

The upside of hunting wild pigs is that you get lots of good-tasting, nutritious meat.

BEAR

Where you'll find them: North America, South America, Europe, Asia.

Bear hunting is common in many parts of the world, but generally only with live ammunition or extremely high-tension professional hunting bows. It's *possible* to catch a bear using a pitfall or spear trap, but this is a *very* dangerous way to get food. A bear can easily kill a human, and a wounded bear will be very aggressive. So although bear meat is an excellent food source, I'd want to be *extremely* sure there was no other food available before I went after a bear in a survival situation. In short, choose your battles carefully – and if you tackle a bear and are ill-equipped, be prepared to lose!

FINDING ANIMALS IN THE ANTARCTIC

The Antarctic is perhaps the most brutal environment on our planet, and the least conducive to life. The temperatures are consistently sub-zero and it's technically a desert because the rainfall levels are so low. The further you head towards the South Pole, the harder it is to survive – not only for you, but for the animals on which you might otherwise rely for food.

However, where the sea meets the land, conditions are a tiny bit friendlier. If you find yourself in a survival situation in these regions, you'll have to do what the great Antarctic explorers of the early twentieth century did, and rely on the continent's two major sources of food. These are seals and penguins. (It's worth remembering that this is what kept Shackleton and his men alive after they'd eaten their dogs – and theirs was among the greatest survival stories ever.)

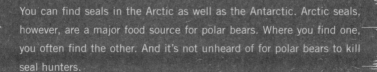

You can find seals in the Arctic as well as the Antarctic. Arctic seals, however, are a major food source for polar bears. Where you find one, you often find the other. And it's not unheard of for polar bears to kill seal hunters.

SEALS

Don't be fooled by a seemingly recumbent seal. Large specimens can be aggressive: they can rear up and attack you. In a survival situation, your best bet is to go after seal pups. They can't swim and can easily be despatched by a sturdy club on the head. (Many are killed like this by seal hunters – you might have seen the tragic footage of this brutal over-hunting, so you really must only consider this to be an option in a genuine survival situation.) Alternatively, you can find a seal hole in the ice and wait for an adult specimen to come up for air before spearing it – but it's a hard way to catch your supper, requiring plenty of patience (and if you've spent a long time without food, time might not be on your side).

There's a lot of good meat on a seal – make sure you cook it first, though, as it can harbour a nasty parasitic worm that causes trichinosis. Don't eat the liver – in common with polar bears (which I don't recommend you think of as a food source, because they're up there with crocs as one of the most dangerous animals in the world) it contains a toxic level of Vitamin A. There is a lot of blubber on a seal, too, which you might at first be tempted to eat for the calories. But be careful: your body will need a lot of water to process that fat. You can also render the fat down to use as fuel.

> Be careful handling seal meat if you've got cuts on your fingers: it can result in a painful and debilitating infection called 'spekk finger'. Nobody quite knows what causes it, but if you get spekk finger you'll need antibiotics if you want to avoid a very nasty illness.

PENGUINS

Penguins are legally protected under the Antarctic treaty, so these are another animal you could only ever consider taking under extreme conditions. Another tough one to catch, because they tend to head immediately for the water if they see any sign of danger. However, they are more likely to stay put when they're nesting.

7

SNARES AND TRAPS

In the previous section, we learned that if animals get even the slightest hint of your presence, they'll be scared off. So logically, the best way to lure them out is not to be there in the first place. That's where snares and traps come in. They can be used in any environment and, as always, they can be improvised using materials you can gather in the field.

We know that for thousands of years our ancestors used snares and traps very much like the ones I'm going to show you. That in itself tells us how effective they are. Our ancestors were meat-eating hunter-gatherers. They lived at a time before modern farming methods made it easier to harvest your own meat, and before guns were available to shoot animals from a distance. Snares and traps, along with some of the killing devices in Chapter 8, were really their only option. And so, if they were going to get their hands on the protein that they required, they had to perfect their snares and traps to the point of maximum efficiency.

It is because of this efficiency that a knowledge of these snares and traps gives you a big advantage in a survival situation. Tracking and hunting can be very exhausting, especially when food – and therefore energy – is scarce. Ensnaring your prey is a lot less labour-intensive, because you simply set the snare and then get on with your other tasks, which means you're not burning precious calories 'hunting'. Moreover, if you've ever seen a rabbit, deer, fox or kangaroo running, you'll know that chasing after it is a fool's errand. If you want to catch it, you'll have to think like our ancestors did: a bit smarter.

A few things to bear in mind:

1. A randomly placed snare won't catch anything. You need to establish that you are in an area where animals pass through. To do this, you need to be proficient in recognizing runs and trails, identifying tracks and droppings, spotting chewed or rubbed vegetation, nesting sites and watering areas. See Chapter 6 for guidance.

2. With a couple of exceptions, the snares and traps I'm going to show you are best for catching small animals. There's really no point setting up a pencil snare to catch a deer, or a bottle trap to catch a wild pig.

3. Most animals will avoid a small snare placed on a wide trail. They'll just walk round it. You can increase your chance of success by making a channel, using branches or foliage on either side of the trail, to 'funnel' your prey towards the snare. Unable to turn left or right, most animals will continue towards the trap (they certainly tend not to walk backwards out of the channel, as they prefer to face the direction in which they're travelling). Alternatively, you can use the multi-noose spring snare on page 150.

4. In a survival situation, you should try to set as many snares as possible – sometimes you need a large number just to catch one animal. Remember: survival is all about playing the odds to your advantage. I always set at least six if I want to catch anything.

5. Remember what we said in the previous section about animals having an excellent sense of smell. Try not to use fresh twigs or branches for your snares (they smell different and unusual, and animals might avoid them). If you handle your snares too much before setting them, your quarry will smell you on them.

So cake your hands in mud first, in order to mask your smell. (Our combat survival instructor in the SAS taught me this and I happily got stuck in – I never needed any encouraging to get muddy!) Alternatively, holding your snare over smoke from a wood fire is another good way of getting rid of your stench. (Animals will avoid fires, but they're generally used to the smell of wood smoke.)

Best of all, if you have any previous kills you can smear the contents of the gall and/or urine bladder over the snare. Just don't use your own urine – it'll make animals run a mile!

The same goes for any material you use to channel a trail.

There are certain smells that will positively *attract* wild animals. You can buy oils of aniseed and rhodium for this purpose. A few drops smeared over your snare will mask your scent and be a beacon to your prey.

6. Camouflage your traps as best you can. Animals aren't stupid. If they see something unfamiliar, they'll avoid it. Try to make your snares blend in with their environment by covering them with light twigs or leaves (but make sure the camouflage doesn't stop them working properly).

7. Your snares must be strong. A caught animal will fight like mad to get free, and any weakness in your construction will be exposed.

8. You must check your snares regularly. Not only is an un-set trap a wasted opportunity in a survival situation, but you need to remember that you're not the only creature out there who might want to eat your catch. And if your animal is not dead in the trap, there's a substantial risk of it breaking free. Finally, if you're going to catch and kill an animal, you want its death to be as fast and pain-free as possible – not to mention that a dead animal will start to decompose very quickly, especially in warm conditions.

9. If your ensnared animal is not dead, approach it with great care. It will fight for its life and can be dangerous, even though it might not seem it. Remember that an infected bite from a tiny rodent can kill you out in the wild . . .

WARNING

Many of these snares and traps may be illegal – it depends on where you are. Always behave lawfully and only ever use them in a genuine and extreme survival situation. Always make sure that you collect or dismantle any unused traps before you leave the vicinity. This stops them causing any unnecessary harm to wild animals who might get caught long after you've split the scene, and in some instances, as you'll see, the traps can be dangerous to unsuspecting humans too.

BAITING

Most of the snares and traps I'm going to show you can be used without bait. However, if you *can* bait them, it will certainly make them more effective.

When considering what bait to use, you should take a few things into consideration. First, take a look around you and see what food source already exists in abundance. Don't use that – you'll never bait a rabbit with a piece of cabbage in the middle of a cabbage field!

Now think about what kind of animal you're trying to trap. If it lives in a tree, you need to bait it with some kind of fruit. If it's a carnivore, you need meat. If it digs in the earth, you should bait it with roots or insects. If it grazes, you need greenery.

You could also try a technique known as 'test baiting'. Before sundown, find yourself an area of flat, dusty earth and drive a few stakes, each 30–40cm in length, into the ground. To each stake, attach some bait. Use lots of different kinds and make a note of what bait you've put where. Go back the following morning to see which items have been taken. Check for tracks in the dusty ground. This will give you a good idea of what animals are in the area, and you can set your traps and bait accordingly.

HOW TO MAKE A NOOSE

Many of the snares below rely on a simple noose, which is very easy to tie. Wire is the best choice for most of these snares, because you can support it in different positions without it losing its shape. So it's a good thing to carry in your survival pack.

If you are using wire, make a small loop at one end by twisting the end several times around itself. Then feed the free end back through this loop (a great tip is to remember that if you do this twice, then the loop won't slacken off once it's tightly drawn).

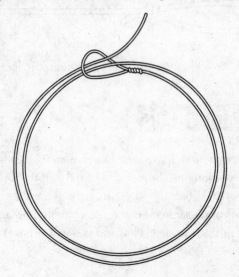

If you are using string or other cordage, first make a slip knot at the end of the string, like this.

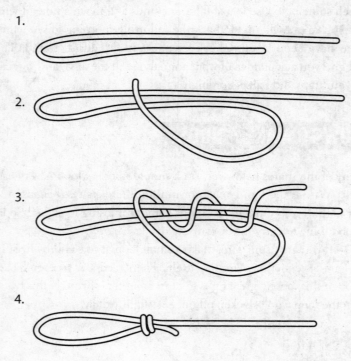

1.

2.

3.

4.

Now feed the end of the string back through the loop.

SNARES AND TRAPS

I categorize snares and traps into four types: they either strangle, dangle, tangle or mangle. These sound brutal, but that's what survival can come down to. Ultimately, these snares and traps are only limited by the scope of your imagination and use of the terrain and resources around you. Think smart, feed well and you will survive!

SIMPLE SNARE (STRANGLE)

The basic idea of a snare is to create a noose that tightens when an animal steps into it. If your quarry knows it's being trapped, it will probably wriggle and try to escape. A good snare, therefore, will grow tighter the more the animal kicks. This won't usually kill the animal – that will be up to you when you find it.

If the noose is made of cordage, there's a chance that it may loosen enough for the animal to escape. Wire is therefore the best choice for a simple snare like this one (it's also a lot harder for a panicked animal to chew through wire than through rope).

This snare is simply a piece of wire with a loop tied at one end and the whole thing fashioned into a noose. If a small animal catches its leg or neck in it and starts to wriggle to escape, the noose will tighten.

You can place a noose like this directly over the mouth of an animal hole (see pages 132–3 for how to check if the animal hole is in use or not). Make sure one end of the snare is firmly tightened to a stake or tree trunk.

Alternatively, you can place this snare on an animal run. In which case, support it off the ground using a couple of small twigs. Make sure that the noose is a good 20–30 cm away from whatever you've fixed it to, otherwise your prey will tend to avoid it.

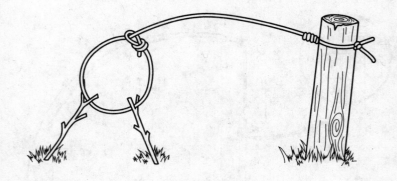

SPRING SNARE (DANGLE)

If you are in an area where there are lots of small, springy saplings emerging from the ground, you can adapt the simple snare on page 149 by setting it under tension. (Alternatively, you can use a low overhanging branch that can be easily pulled down but will snap back up when released. This is a very useful trick to have up your sleeve. A spring snare raises your prey up off the ground where it is (a) less likely to escape and (b) out of the reach of other predators.

To make a spring snare, select a site where a springy sapling is close to an animal run. Remove the sapling's branches and cut off the top. Now find yourself a sturdy stake that you can drive into the ground just below the point where the sapling will reach when you bend it by 90 degrees. Cut a notch in the stake. Use your knife to fashion a hook out of a piece of wood. Tie it to a piece of cordage, then attach the other end of the cord to the top of the sapling. Bend the sapling over, and engage the hook into the notch in your stake, as shown in the picture below.

Now attach a simple snare to the hook and set it raised from the ground by two small twigs. The idea is that when your prey becomes ensnared and starts to wriggle, the movement will cause the hook to become disengaged. The sapling will straighten up,

pulling your prey up from the ground and leaving it dangling in the air.

Once you get the hang of the spring snare, you can increase your chances of a catch by baiting it. To do this, you need to tie another piece of cordage to a small stake that you insert only lightly into the ground (because it has to fly up again when the snare is sprung). Tie the free end of your cordage to the hook and fit your bait so that it is just above the noose of the snare. This is a particularly effective way of catching rabbits – a bit of apple is good as bait, as are some rabbit droppings (they tend to attract other rabbits). You could also use this baited spring snare to catch a fox – any bit of old meat will attract them, the smellier the better.

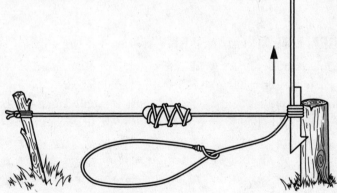

TREE SNARE (TANGLE)

This is a particularly good way of catching squirrels and other tree-climbing animals. It works on the principle that these creatures will always take the easiest route up a tree. Many small trees lean one way or another. If you examine the bark on the upper side of the lean, you might find scratches where small animals have climbed up the trunk. A tree like that would be a very good candidate for a tree snare.

Your first job is to find a pole that you can lean up against the tree. Your average squirrel will be more likely to use this pole because it's a simpler method of getting where he wants to go. Now tie some simple snares – wire is best, because it holds its shape better, but the nooses only need to be about 2.5–5cm in diameter to catch squirrels – along the pole in the path of the animal. The more snares you put along the pole, the higher the chance of your prey becoming tangled.

cross section of pole and wire

LOGFALL TRAP (MANGLE)

A logfall trap uses heavy weights to crush your prey. You need to be extra careful setting a trap like this – the logs don't care if they fall on animals or humans, and they can do just as much damage to both! This is probably best if there are two or more of you to help in its construction.

First of all, find yourself a series of long, straight logs – the heavier the better (if they're not heavy, you can cover the trap with stones). Bind them together as if you're making a raft, with two logs set at right angles at the top and bottom for support.

You now need to make what's called a 'figure 4' trigger – this is hard to describe, but the picture on the right should show you how it works. It is designed to hold the trap up, but to collapse when an animal comes along and takes the bait – which you should spear on to the free end of the horizontal part of the four.

PITFALL TRAP (MANGLE)

The idea of a pitfall trap is very simple. You dig a hole deep enough to catch your target prey – make sure it's directly in the middle of a trail that you know the animal is likely to walk along – then cover it with a network of branches and dried leaves or moss. To make it more lethal, you can add a 'punji stick' to the bottom – a sharpened stick, pointing upwards.

I'll be honest with you: pitfall traps have a lot of downsides. You'll have to expend a huge amount of energy digging the pit in the first place, and most game will tend to avoid the unfamiliar ground that you're presenting them with anyway. But it's a good trick to have up your sleeve if you've got no other way of catching food. Just use your common sense – never leave pitfall traps covered when you leave the area. Even better, fill them back in again when you're done.

DROWNING SNARE (TANGLE)

If you're hunting near deep water with a steep bank, a drowning snare could be your best bet. It's good because it not only catches the animal, it also despatches it quickly by drowning. If the water is cold, it will keep the carcass fresher for longer, and it will be safe from scavengers.

To make a drowning snare, tie the loose end of your snare to a large boulder. With a separate, long piece of cordage or wire, tie a short, stubby stick to the noose. Place the snare near the water where you think an animal will come to drink, and balance the rock on the edge of the steep bank. When your prey becomes entangled, the rock will fall into the water (assuming it's balanced very precariously and the animal is large enough), taking the animal with it. The stick will act as a float to tell you whereabouts under the water your dinner is.

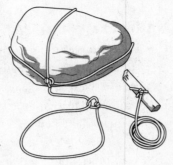

BOTTLE TRAP

This is a good trap for catching small rodents, particularly mice and voles. It's called the bottle trap because you have to make a hole in the ground, about 30–40cm deep, the same shape as a bottle – wider at the bottom than at the top. Place a flat stone or a piece of bark over it, with a couple of small stones to keep the larger stone or bark 2–3cm off the ground. Mice or voles will find this an attractive place to hide from other predators, but when they fall into the hole they won't be able to escape because of its shape.

Be warned: in the right terrain, a bottle-trap hole might also attract snakes. Don't go sticking your hand in there unless you're sure it's safe.

TRAPPING BIRDS

On pages 123–4 you'll see a technique for catching sea birds. This works for land birds too, but there are a number of other methods you can employ for catching this abundant food source. In fact, birds can often be easier to catch than mammals, and it's worth remembering that all species of bird are edible (though they don't all taste good – vulture meat is particularly foul).

Again, remember that these are only ever to be used in a genuine survival situation.

OJIBWA BIRD POLE

This is a very ancient, primitive trapping device. But effective. It works in wide, open spaces, using the principle that birds will naturally seek a perch on which to land. If there are no natural perches nearby, they'll go for the one with which you provide them.

First, find a straight pole a couple of metres long. Strip it of any branches or foliage, then use your knife to make a sharp point at one end. Drill a small hole in the other end (the point of a knife can work for this, if you don't have an awl in your pack). Drive the sharp end into the ground until it's secure.

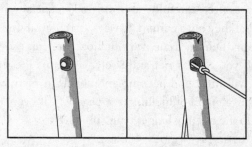

Now find a stick, about 10–15cm long, that will loosely fit into the hole. This will be your perch. Don't put it in place just yet.

Tie a rock to a piece of thin cordage. The rock needs to be about the same weight as the bird you're going to catch (if it's too heavy, your snare will just cut the bird's feet off). Thread the loose end through the hole, then make a simple knot so that it's next to the hole when the rock is lifted from the ground.

Insert the stick into the hole so that the knot you've made holds it loosely in position.

Now make a noose with the loose end of the cord. Drape it over the perch as shown.

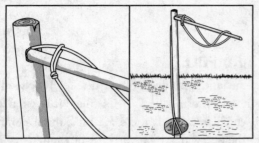

When a bird lands on the perch, it will dislodge it. The rock will fall and the noose will tighten around the bird's feet.

BIRD NET (TANGLE)

If you observe birds carefully, and over a decent amount of time, you'll find that they have certain 'flyways' that they always use. When you're hunting them in a survival situation, you can use this to your advantage. If you have a net and stretch it across the flyway, some of the birds will hit it and become entangled. In many ways, this is the avian equivalent of gillnetting (see page 92). It's very effective, so don't keep up the net any longer than you have to.

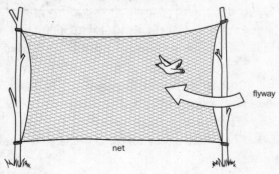

flyway

net

STALKING FRESHWATER BIRDS

If you're near a pond, lake or river where birds have settled, it is possible to catch them by hand. You just need to get close enough, and to do this you need to camouflage yourself. Strap reeds, foliage or whatever local vegetation you can find around your head. Then get into the water and slowly approach your prey, moving with the current if there is one, and with only your camouflaged head above the water. If a bird allows you to get close enough, grab its feet and pull it down underneath the water, then break its neck while it is submerged. But be careful: birds can become very aggressive when they're under attack, and some of them are surprisingly strong.

8

OTHER KILLING DEVICES

The snares and traps that you saw in the previous section are all excellent ways of catching small game and birds when you require food in a survival situation. But sometimes you need to take a more hands-on approach. If, for example, you know there are wild pigs or deer in the vicinity, and your snares or traps aren't working for these larger game animals, it might be that you have to stalk them and do the dirty work yourself if you're going to eat and survive. To do this, you're going to need a weapon of some kind. Here are some ideas.

KNIFE

In Norway they say: 'A knife-less man is a life-less man.' That's how much importance traditional Norwegian hunters place on having a good blade with them at all times. If you could only take one thing with you into the wild, you'd have to give serious consideration to making that one thing a good, sharp knife. In the field of wild food survival, it's crucial.

You'll need a knife to make many of the devices listed in this chapter – and it is, of course, a killing device all on its own. You'll want a knife to harvest and prepare wild plants or prepare freshly caught fish. In the pages that follow, you'll see that it's crucial for preparing any wild game that you've managed to snare, catch and kill.

I wouldn't ever venture out into the wild without a good knife.

And in a survival situation, your chances of making it are so much greater if you have one.

In addition, I once heard an old huntsman say that 'one knife is no knife and two knives are one'. Which means that it is always best to have a back-up in case a blade breaks or you lose your primary knife. It really just reinforces how much a knife can do to help you survive and hunt in the wild. So prepare well and don't compromise on a good blade . . . or two.

Here are a few things you need to consider when you choose a knife.

Tang

The 'tang' of a blade is how much of the metal extends into the handle. A 'full tang' blade is the best: the metal extends all the way through the handle, which makes for a much stronger and safer knife.

Handle

You need a good, solid handle (if it's hollow, it's not a full-tang knife). It needs a good grip and to feel comfortable in the hand.

Metal

Your options are stainless steel or carbon steel. Stainless steel is a lot more rugged and won't rust, but takes a bit more sharpening. Carbon steel is easier to sharpen, but will tend to rust very easily.

Blade shape

Smooth or serrated? A smooth blade is much easier to sharpen and will achieve pretty much everything you need it to do in this book. A serrated blade will allow you to saw through tougher items. A good compromise is a mainly smooth blade with a small serrated area near the handle. Think, too, about the size of your blade: a very big blade might seem like a good idea, but it's more difficult to carry and can be cumbersome to use.

Sheath

Really important. It serves two purposes: it protects you, and the people around you, from the blade, and it protects the blade from the elements. If you're not using your knife, you must always sheathe it and stow it away carefully.

The BG Gerber Ultimate Pro Knife that I have designed specifically for survival has a full-tang fixed blade, 12.2cm (4.8 inches) in length. The blade is made from high-quality stainless steel. The handle is shaped to fit comfortably in the palm and is rubberized for a good grip. It comes with a good, sturdy sheath from which the blade won't accidentally escape, and it has a built-in sharpener that can make a critical difference in the wild. But the truth is that any blade, as long as it is full tang, will do the trick. A knife is only as good as its operator. I choose to use a blade that I have designed and that I know delivers when I need it to. But you can decide for yourself, and we all have favourites.

It doesn't matter how good your knife is if it's blunt. A blunt knife will mean you botch the job of killing a small animal, or of skinning and gutting your quarry. In a survival situation you can't afford to do that. It is also true that a blunt knife is more dangerous than a razor-sharp one, as you then have to press much harder. Hence the chances of a blade slipping or jumping uncontrollably in your hand (or even worse, near the artery running down the inside of your leg, which people often use to support whatever they're cutting) are much increased. A very sharp knife, on the other hand, can be handled gently with minimum force or strain so any slippage is unlikely. People are often surprised when I tell them to sharpen their blade for maximum safety!

So, always make sure your knife is sharp before setting out into the wild – you can find full instructions on how to do this in my book *Living Wild*. If you find yourself in the field with a blunt blade and no sharpening equipment, you can make a pretty good job of it by sharpening the blade against a smooth, flat stone – preferably wet, so a stone in a river bed or stream is good. If you're right-handed, hold the handle in your right hand and place the fingertips of your left hand against the blade. Make an angle of about 10 degrees between the knife and the stone. Press slightly, then sharpen by pushing *away* from you in a slight clockwise motion. Now flip the knife and do the other side, but this time sharpen in an anticlockwise motion.

BOW AND ARROW

When it's well made, the bow and arrow is one of the ultimate hunting devices. It has advantages even over a gun, because it's quiet, light and can easily be constructed from scratch. Indigenous people have been using bows and arrows for thousands of years because they allow you to stalk your prey at a distance and then despatch them – or at least wound them – with a very sharp, very fast, very silent projectile. If it's good enough for them, it's good enough for us.

Making a bow is an art form. Master bow-makers spend years perfecting it and many months making each bow. Obviously, when you're hungry and out in the field you don't have that luxury. So what follows are instructions on how to make a serviceable bow and arrow for use in a survival situation.

Be warned that in many parts of the world (including the UK) hunting with a bow is illegal, and with good reason. Inexpertly handled, one of these things could easily kill more than just an animal.

BOW

To make your bow, you need to find a long, straight pole, just over a metre in length. (The longer the pole – or 'stave' – the more difficult the bow is to handle.) You can make a bow out of pretty much any kind of wood. The best is yew, but that's rare so you should just do the best with what you have. Whatever wood you use, you should try to do three things. The first is to remove the bark. The second is to taper the two ends to make it more pliable when you pull back the bowstring (see picture – the ends should be about 1.5–2cm in thickness, the centre about 2.5–3cm). The third is to dry it out a little by placing it over your fire, not too close to the flame: if you can reduce the moisture content of your bow, it will last longer.

Now you need to cut little notches at both ends of the bow, as shown. This will enable you to tie your bowstring without it slipping.

You'll need a length of string or natural cordage (see Appendix A) for your bowstring. Tie it round the notches of your bow so that there is only the slightest amount of tension in the string.

Now it's time to think about your arrows.

ARROW

There are three components to consider: the shaft, the arrowhead and the feathering.

To make the shaft, you need long, straight branches about 7.5mm in diameter. It's quite hard to find ones that are *totally* straight, but try to ensure they don't have any noticeable kinks; the best place to look is in dense forest where branches are struggling up towards the sun. Cut the branches long – about 60–70cm – because you can always trim them down later if you need to. Now remove the bark and dry them over your fire as you did with the bow. The more moisture you can expel, the stiffer they will be and the better they'll fly. If there are any slight bends in the wood, you can straighten them out by bending them back over the fire for 15–20 seconds. Now cut a notch at one end – this will be where your bowstring slots in.

For the arrow to be an effective weapon, it will need a point. You can do this by whittling one end away with your knife until it's razor sharp, then hardening it off in the fire. But as a killing device, it will be deadlier with a harder tip. I've been in places where I've found ancient arrow tips on the ground, left over from the days when indigenous people used bows and arrows as their principal hunting weapons. But most of the time you'll have to fashion them yourself.

The usual materials for making arrowheads in the field are flint and bone. Flint can be chipped away to make a sharp point; likewise, bone can be ground against a rough stone.

Once you have your arrowhead, you need to fix it to the shaft. A good way of doing this is to split the end of the shaft, force the arrowhead into the split and then bind it tightly with cordage.

If you want to make the arrowhead even more secure, and you have the time and the materials to hand, you can make pine-pitch glue. To do this, you need to extract some sap from a pine – or any other coniferous tree – by hacking a small wound into the tree about a metre from the base. Peel away the bark and strap a collecting vessel just below the wound. The sap should trickle in. Now take a bit of (cold) charcoal from a recent fire and grind it to a powder between a couple of rocks. Light yourself a small fire and warm the pine resin through, before adding some crushed charcoal (1 part charcoal to 5 parts resin). Now, heat your arrowhead over the fire and smear it with your pine-pitch glue, before fitting it to the shaft and proceeding as above.

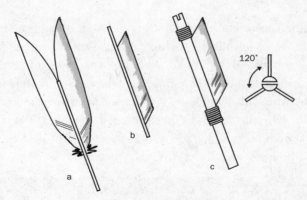

Finally, you can improve the accuracy of your arrows by feathering, or 'fletching', them. Collect some decent-sized feathers – the sturdier the better. Use your knife to split them in half along the length of the quill. Ideally, you'll need three of these half-feathers, which you should tie to the end of your arrow so they are evenly spaced at angles of 120 degrees.

USING YOUR BOW AND ARROW

Whatever you do, don't try to use your new weapon for the first time when a wild pig is advancing. You need to practise with it first.

Fit the notch at the end of the arrow into the bowstring. Extend your bow hand fully (if you're right-handed, this will be your left hand, and vice versa), with your hand just below the tip of the arrow. Now pull back the string as hard as you can. Keep the tip of the arrow at eye level and aim carefully at your target. To fire, simply let go with your arrow hand.

When using your bow and arrow for hunting, you need to get in as close to your prey as possible. Aim just behind the top of the front legs, where its vital organs are. This gives you the best chance of success. Even if you miss the actual heart, you will puncture either the lungs or an artery and this will soon kill the animal. Arrows will cut through both arteries and veins, resulting in massive haemorrhaging.

In addition to severe bleeding, arrows passing through both lungs cause the lungs to collapse, leading to rapid death due to suffocation.

If you wound the animal, it will probably try to run. But it will be losing blood and will soon weaken. Take another shot if you can, but otherwise stalk it and finish it off humanely when it's too weak to get away. This might sound a bit grim, but the realities of survival are often far from pretty and, if you want to make it in the wild, this is the kind of thing you sometimes have to be prepared to do.

Once you've gone to the trouble of making your bow and arrow, you can preserve their useful life by oiling the wood. In the field, this can be done by rendering down some fat from an animal you've caught and rubbing this over the wood. You'll find that this makes the bow keep its springiness for longer.

SPEAR

We've already discussed using a spear for fishing, but you can also use it for hunting small game. Think of it as a large, hand-held arrow – the process for making the tip is the same (see above). However, when you're cutting your pole to make a spear, think about how you're going to use it: is it for jabbing or throwing? Jabbing spears need to be longer – up to 2 metres. Throwing spears should be about half that.

A word of warning: please don't be tempted to lash your knife to your spear. In the wild, and in any survival situation, your knife is one of your most precious assets. Using it at the end of a spear for hunting (and especially for spearfishing) is a sure-fire way to ruin it. (The only exception to this rule is if you need to make a spear for self-defence against aggressive predators.)

CATAPULT

You might think of a catapult as being a schoolboy's toy, but it's much more than that. A strong, well-made catapult can deliver a lot more energy to a target than an air rifle. In fact, lead shot is a very effective projectile if you have some with you. But, whatever you use, in the right hands it will be easily powerful enough to kill small game such as rabbits and squirrels. It has the advantage of being extremely lightweight and portable. Plus, of course, you can improvise one.

First, you need a sturdy, forked stick. Try to find one where the arms have the same diameter, and where the handle is thick enough to grip firmly. If the wood has a little bit of give, so much the better: hazel, ash and hawthorn are all good choices. Peel off the bark and scrape away any bumps that make it uncomfortable to hold. Now carve a notch round the circumference of each of the arms a couple of centimetres from the end.

Next, you need a piece of elastic material. A strip of latex rubber about a metre long is ideal, but you could use all sorts of things: washing-up gloves, bicycle inner tubes, elastic bands . . . (If you don't have anything elastic, you can adapt your catapult to become a slingshot. It worked for David against Goliath!) You'll also need something to act as a pouch to hold your ammunition as you're pulling the catapult back. A strip of leather is best – in a survival situation you could cut off the tongue of your boot and use that. Finally you'll need some cordage. Fishing twine is ideal.

Make a couple of holes, one on either side of your leather pouch, and thread your elastic material through that. Centre the pouch. Now tie each end of the elastic to the arms of your catapult, and secure them in place with your fishing twine, which you need to wrap many times round the notches you've cut in the arms.

Now: practise. Set up a target about 10 metres away (an empty drinks can is good) and arm yourself with a stash of small projectiles: small stones, ball bearings, lead shot . . . If you're right-handed, hold the catapult in your left hand and pull the projectile back with your right, holding it between your thumb and index finger. (Swap this round if you're left-handed.) Straighten your left arm, pull back the pouch, aim and fire. Make sure you hold the catapult straight or you risk hitting your hand, which really hurts. Once you can reliably hit your target at 10 metres, you're ready to go hunting.

Don't underestimate the power of your catapult. It will easily kill small game, and can do the same to humans. Think of it like a gun and don't point it at anyone.

RABBIT STICK

A rabbit stick – also simply called a throwing stick – is one of the earliest survival tools. It's also one of the most effective, especially on open grassland where it can fly for over 100 metres just a few centimetres above the ground, and it is also one of the easiest to improvise in the wild.

Think of a rabbit stick as a boomerang that doesn't come back. It's little more than a curved branch with a slight bend in it. Sounds simple, but it's deadly in the right hands. People spend a lot of time carving rabbit sticks to their own specification, but in a survival situation you can use any curved stick between 30cm and 60cm in length and about as thick as your wrist. The smaller the stick, the faster it will fly; but a bigger stick will be more accurate. You can sharpen the ends to make it more deadly.

As with all these devices, you will need to practise before you are adept at using them. The rabbit stick can be thrown in three different ways:

1. Diagonally. Release the stick up and away from your body at a 45-degree angle. This is the most accurate way of throwing it, and is good for wide, open spaces.

2. **Vertically.** Throw the stick over-arm so it rotates vertically end over end. This is a less accurate way of throwing, but it allows the stick to pass through vertically growing vegetation, like trees.

3. **Horizontally.** Throw the stick sideways, so it's at 90 degrees to your body. This is a good way of hitting birds floating on open water.

A well-thrown rabbit stick can efficiently kill small game. At the very least, a direct hit will stun your prey so you can approach and finish off the job quickly and cleanly.

GUN

By gun, I mean an air rifle. That's the only gun you're allowed to carry unlicensed in the UK, where I live, and the same goes for many other parts of the world too.

An air rifle is a lot safer than a live-ammo weapon, but let's get one thing straight: it's still not a toy. It might not be as deadly as a shotgun, but it can still inflict a lot of damage and needs to be treated, and used, with respect. The same goes for all the killing devices we've already discussed. (You'll often find that the use of an air rifle in the field is legal, whereas the use of an improvised bow and arrow is not. It's up to you to find out what the rules are where you live.)

In a survival situation, you have to use whatever the most expedient tools are to hand. Air-rifle pellets are extremely lightweight, so it's easy to carry a stash of ready ammunition with you. Many types of air gun are almost silent shooters, so you can hunt without scaring away every animal in the vicinity. All in all, they're an extremely effective way of killing all manner of small game wherever you find yourself in the wild. So if you have an air rifle with you, you'd be foolish not to use it when the situation demands.

There are, however, other good reasons for *learning* how to use an air rifle. Although it's easier to learn how to use it more effectively than, say, a rabbit stick, possession of a firearm by no means guarantees that you're going to end up with food in your belly. To use an air rifle properly, you'll need to become extremely proficient in the age-old techniques of tracking and concealment. Having the skill set necessary to do this can only help you in a survival situation. You should consider an air-rifle awareness training course, but here are the basics.

TYPES OF AIR RIFLE

Air rifles fall into many different categories. The most common types for hunting are compressed-air-powered rifles and spring-powered rifles.

Compressed-air-powered rifles use a cartridge of compressed carbon dioxide to fire the pellet. A 12g cartridge will give you about fifty shots. The advantage of using one of these is that you don't have to pump or spring-load your weapon before taking a shot. On the down side, they're not quite as powerful as other types of rifle.

Spring-powered rifles require you to pull back a lever to compress the air in the barrel. This means that you need to cock the rifle each time you take a shot, and you also tend to get a bit of a recoil when you fire, which can make them a little less accurate. But on the plus side they tend to be more powerful than compressed-air rifles.

TYPES OF PELLET

If you're going to use an air rifle, you need to make a decision about what kind of pellet you're going to use. Your choice will probably boil down to a .22-calibre or a .177-calibre. A .22 is slightly bigger, so will kill your target more efficiently. The .177-calibre, being lighter, is slightly easier to shoot accurately. But, with practice, you should find either up to the job.

WHAT TO SHOOT

Air guns are not as powerful as shotguns. You won't be able to use them efficiently to kill anything much larger than a rabbit or a game bird. Trying to kill a deer-sized animal with an air rifle is likely to be a waste of your time and to cause unnecessary suffering to the animal itself. Don't do it.

PRACTICE MAKES PERFECT

If you're going to hunt with an air rifle, you need to become proficient with it before you go out in the field. You'll need to make sure that your sights are properly zeroed, and that you can accurately hit a target about the size of a five-pence piece at about 30–35 metres. That's the kill zone on a rabbit-sized animal. In order to make a clean kill, you need to be able to guarantee a head shot first time. And the best way to learn how to do that is to get out in the field with someone who is expert in the handling of an air rifle.

That doesn't mean, however, that you can't practise on your own – because you don't even need a gun in your hand to master the most important skills required to get a bird or small animal in your sights. Those skills are stillness, stealth and silence. Get out into the field and search for a trail or a roosting spot. Get up close. Camouflage yourself as best you can. Now stay still. *Really* still. The best way not to be seen by an animal is not to move a muscle. They are unbelievably sensitive to movement, but if you remain statue-still, you've a chance of going unobserved. If you want to know a bit more about how to remain unseen in the field, see my book *Living Wild*.

9

PREPARING AND COOKING YOUR KILL

So you've managed to trap, catch or shoot an animal. Positive work so far towards a life-giving meal for you. But your work's not done yet. In order to get that precious fuel inside your stomach, you need to skin, butcher and prepare your animal for cooking. And, in a survival situation, you need to make sure that no scrap of useable food is wasted.

What's more, you need to do it quickly. Freshly killed animals can become inedible very soon after they're killed, especially in warm climates. This means that you should carry out the act of 'field dressing' the animal – i.e. getting rid of the soft organs and blood – as soon as possible. It's a dirty job, but someone's got to do it, and if you are squeamish, this is a good time to get over it. It is why I find it best to get mucky early, so then you don't get too precious about trying to stay 'sanitized' later on. (The better you are at butchering, the less blood you should need to get on you. But however good you are, you will inevitably get bloody at some stage, so get used to it early.)

Then you need to know how to skin and possibly joint your kill before preparing it for cooking.

In this section, we're going to start off with small game. Think rabbits, hares, otters, squirrels – the techniques are similar for all of them, and these are what you're most likely to be catching. (Large game are harder to hunt, and in a survival situation you'll expend a lot of energy trying to do so.) Moreover, it's best to start with small game when you're learning how to butcher animals in the wild, because they're easier to handle and you'll see that some of the techniques hold for larger game too. Finally, we'll look at birds and how to prepare them for the pot.

SMALL ANIMALS

THE BEST WAY TO KILL SMALL ANIMALS

If you've snared an animal for food, you owe it to the beast to give it a quick, painless death. There are a number of ways of doing this.

The first is to break its neck. This is a particularly efficient way of despatching rabbits. With one hand, hold them firmly by the hind legs. Now grip the animal's neck with your other hand and with a strong, sudden movement, yank back the legs and both pull and twist the neck so that it breaks.

By this time, your animal should be dead (even if it is still twitching a bit). If you're in any doubt, you can employ an alternative method of despatch: a blow to the head. Hold the animal upside down and give it a few solid blows on the back of its head. You can use a karate chop action to do this if you like, or use a stout, heavy stick. Both methods will work if done firmly and swiftly. Get it right, though, and make sure you minimize any suffering.

The third way to despatch your prey is to stab it. It's messier, but sometimes necessary, especially if your animal is panicked, struggling and there's a risk it might bite you. You can stab the animal from a distance with a long, sturdy, sharpened stick. Or, if you can get in close enough, stab it at the back of the neck. Alternatively, if it is a small animal, quickly remove the whole head with your knife (you might want to hold it down under your coat or some other piece of protective material while you do this, to protect your hands from a frightened, biting animal).

FIELD DRESSING SMALL ANIMALS
Bleeding

There's no doubt that you'll get cleaner, tastier meat if you drain off an animal's blood before butchering it. But that's not the only reason we do it. All mammals are warm-blooded, and that blood stays warm for a while after the animal is dead. But if we want our catch to stay fresh, it needs to be as cool as possible as quickly as possible.

To bleed an animal, you need to sever its jugular and/or carotid artery. They can be difficult to find, so it's often simplest either to cut its throat, ear to ear, or to remove the entire head. Suspend it upside down from a tree branch first, though you should make sure to tie it by a leg joint rather than simply by a paw so it doesn't slip.

There are plenty of indigenous peoples who regularly drink the blood of a freshly killed animal (Inuits believe seal blood gives them strength, and the Maasai of Tanzania tap blood from a live animal and mix it with milk, before letting the wound heal). And the history of survival exploits is full of stories of people drinking blood to stay alive. In fact I have drunk it many times, both with the Tuva tribe in Siberia and the Sami in northern Norway. Never very nice, but the blood is warm, salty and the ultimate survival food.

In extreme situations, the freshly drawn blood of a newly killed animal is a rich source of liquid and nutrients. If you find yourself in such a position, you can catch the blood of your prey in a container and drink it. But be wary. Although fresh, raw blood can be edible – I have never suffered any ill effects when I have drunk it in the wild straight from a freshly killed animal – it *is* possible to catch diseases from it, so you should do it in the field only if the alternative is much worse. Remember: survival is all about weighing up risks. (Also remember that if you cut the animal's oesophagus, the contents of its stomach might drain out as you're bleeding it and contaminate the blood. Not good. So cut the throat carefully, and not too deeply.)

Bleeding an animal can be a messy business. If you can, clean the blood off the carcass in a stream once you've done it.

If you've set up camp, all field dressing, but especially bleeding, should be done at a good distance from the camp itself. Fresh blood and animal parts will attract other scavengers. You really don't want your living and sleeping quarters to be overrun by them. While we're on the subject, don't let anyone tell you that burying animal parts will stop other scavengers from smelling them – it won't. If you're intending to trap more animals, it's a good idea to field dress them in the vicinity of the traps to attract more animals to the area. If you must do it in your camp, it's best to dispose of any remains by burning them in the fire.

Skinning

Skinning a small animal is a lot easier than it looks. If you can, cut off all four feet just above the knees. Now, make a small incision in the hide (the easiest place to do it is around the stomach, at right angles to the length of the body). Be careful with this cut. You want to pierce the body sufficiently to pull away the skin, but not so much that you break into the organs and soft tissue sac.

Once you've made the cut, insert one finger from each hand into the incision and peel the skin back. You should find that it comes away quite easily, though this will depend on the type and age of the animal: the skin will pull away from a young rabbit without much difficulty, for example, but an old hare or squirrel will take a bit more effort.

You don't always need to skin small animals. You can just lie them on the edge of a burning fire. This will burn away the fur and you'll end up with a very charred-looking dinner. But the skin will have protected the meat, which should be nicely cooked through. When you pierce the skin with a knife and the juices run clear, it's good to gut and eat. I've done this with all sorts of animals, from skunks to possums!

Gutting

As with bleeding, you want to gut your animal as soon as possible. The contents of its intestine, stomach and bladder can all start to turn rancid very quickly.

Turn your animal belly up and locate the urinary tract just between its hind legs. With a sharp knife, make an incision deep enough to pierce the skin, but not so deep that you puncture the intestine and other innards. Work your knife up through the skin until you reach the breast bone. You can now peel open the underside of the animal to reveal the stomach, intestines and other organs.

Don't be tempted just to scoop everything out. Although you're going to discard most of the insides, there are edible parts here. There will be a membrane that connects the stomach to the kidneys. Carefully separate this so the kidneys are still attached to the body. Remove the intestines and the bladder (if it splits and you get urine over the meat, make sure you wash it before cooking). The lungs need to come out, but you should keep the heart and liver as these are good eating.

You'll definitely want to eat the liver, kidneys and heart – they are high-value foods, full of nutrients – but you can use bits of the remaining offal to bait snares, or as fish bait. Let the offal get a bit smelly and it will make particularly good alligator bait (see page 242).

Once you've prepared your animal, it's best to get it cooked or preserved (see Chapter 10) as quickly as possible. But if you're not going to eat it immediately, suspend it somewhere off the ground so scavengers can't get at it. You haven't gone to all this trouble to provide dinner for someone else!

Pay attention to the liver. It should be smooth, wet and a deep red colour. If it isn't, or if it has white spots on it, it could be the sign of a diseased animal. That doesn't necessarily mean you can't eat the animal's flesh (you'll have to make that call depending on your situation), but you should discard the liver. It's also worth checking the heart and kidneys for signs of worms or other parasites. If you find 'em, chuck 'em.

COOKING SMALL ANIMALS

Once you've caught and prepared your small game, it's time to cook.

Stewing

If you have a pot, stewing is the best way to get the most nutrition out of your meat. All the fats and minerals leach out into the cooking liquor, which you'll end up drinking. That way they're not wasted: you get food, nutrients and liquid all at the same time.

How you stew an animal in the wild depends entirely on what ingredients you have to hand. As with so many things in the field, you've just got to improvise. This is simply a guideline.

Cut your animal into small bits (don't worry too much about jointing it neatly). Now think about what else you might be able to add to the stew. If you have any grains in your store – think barley, oats, rice – they will massively enrich the stew, so add them now, along with any spices you might have in your trail spice pack. Add water to cover and place over your fire. Let your stew simmer slowly. Stir it with a spoon or a clean stick to stop the contents sticking to the bottom of the pot. When the grains are almost cooked, add whatever vegetables you might have to hand, or any other edible wild plants to flavour the stew.

When everything is cooked through, let it cool a little, then remove the animal pieces. Strip the meat from the bones and return it to the pot. You can discard the cooked bones now – but not near your camp.

Even if you don't have the luxury of extra flavourings like grains or wild plants, this is still a good way of cooking your kill to get the maximum food value out of it.

Spit-roasting

If you don't have any cooking pots, spit-roasting your small animal is the way to go.

You can make a simple spit by inserting two sticks, each with a forked end, either side of your fire, then resting a skewer across them. When you spit-roast your small animal there are a couple of problems you're likely to encounter. The first is securing the carcass to the skewer in such a way that, when you rotate the spit, you also rotate the carcass. Second, because the weight of the spit won't be evenly distributed, it will want to spin to one position all the time.

You can deal with both these problems by choosing your skewer carefully. Find a green branch (so it doesn't burn through easily) with at least two decent side shoots and several shoots at the end. If you can, soak it in water to make it more fire resistant. Skewer your animal in such a way that you can use the side shoots to pierce the carcass and hold it in place. You can now use the shoots at the end to hold the spit in place by means of a vertical stick stuck into the ground, as shown. When you want to rotate the spit, simply adjust the position of these end shoots so the vertical stick keeps them in place.

SOME RECIPES FOR COOKING SMALL ANIMALS IN THE WILD

Sometimes in the wild you don't have the luxury of doing anything but get your meat cooked and eaten, quickly. Sometimes you might have more time, or the occasional extra ingredient to play with. Half the battle of good survival is keeping your morale up, and a decent meal will go a long way towards doing that. With that in mind, here are a few simple recipes you might try with your freshly killed catch.

Squirrel Soup

Squirrels provide one of the sweetest wild meats out there, but some of the gaminess can be removed by soaking it in salty water.

To make this soup, first skin and gut your squirrel, removing its head and feet. Put it in a pot and just cover it with water, to which you've added half a teaspoon of salt. Soak overnight, or at least for a few hours. Bring the pot to the boil and simmer for 10 minutes, then discard the water, cover the squirrel with fresh water and boil it for a couple of hours. When cool enough to handle, remove the squirrel, reserving the liquid. Strip the meat from the squirrel and return it to the liquid with whatever vegetables you have to hand. An onion and a couple of potatoes would be awesome. A carrot or some beans and you're in heaven. Boil again until the vegetables are cooked through, then season with salt, pepper and whatever else you fancy from your trail spice mix.

Clay-roasted Porcupine

Porcupines are a lot tastier than you might think. If you're in porcupine territory (Asia, Europe, Africa, North and South America) they can be easy to catch because they don't move very quickly. You do, however, have to be careful of the quills, which can easily pierce your skin and cause a nasty wound – a potential problem if you're stalking it in a tightly enclosed area, as I once did when I hunted porcupine with the San bushmen in Namibia. A whack on the head with a stout stick is enough to kill them, but then you have the issue of getting past the spines. It's not too much of a problem, though. The porcupine has no spines on the underside of its belly, so if you roll it over on to its back you can skin it in the same way that you skin a rabbit. The San bushmen simply laid hunks of porcupine meat directly on to the embers of their fire to cook it through.

Another way of removing the spines is to place the whole animal on a fire just long enough that the spines burn away.

Alternatively, you can clay roast it. Follow the instructions on page 188 for pit-roasting a joint from a large animal, but first smear the gutted porcupine in a thick layer of clay or wet mud, so the spines are completely encased in a good few centimetres. Place the porcupine in the oven pit, cover in the usual way and let it cook for 3–4 hours. When you remove it from the pit and crack it open, the spines should come away with the baked mud, leaving you with some good, tender meat.

Braised Beaver Tail

Beaver meat can be very tasty – rich, dark, not unlike venison. You can cook it in all the usual ways – it's particularly good roasted over the camp-fire – and it's worth paying attention to the tail, where there is a large amount of good-quality meat. If you carefully skin the tail and cut the meat into 2.5cm cubes, you've got the basis of a really fantastic meal. Put your cubed beaver meat in a pot with some spices and, if you have them, some sliced onion, a few handfuls of beans and some chopped garlic (leaves of wild garlic would do just fine – see page 43). Simmer until the beans are tender, then season and eat.

LARGE ANIMALS

Larger animals you might find yourself hunting include deer and its relatives (elk, moose, reindeer . . .), antelope, camels, wild pig, bison, buffalo, crocodiles and large monitor lizards (more on those later) – even bears if you're in bear country (though, as I've already said, you'd have to be pretty hungry before going after my namesake, because bears are smart, aggressive and strong, and definitely can be deadly).

Hunting large animals in a survival situation is a high-stakes game. You're likely to expend a lot more energy, but you get a lot more food in return. Here's what to do if you manage it.

IS YOUR ANIMAL DEAD?

If you've hunted your animal with a gun or one of the improvised killing devices, you still need to make sure it's dead before you get too close. An animal's eyes being closed does not necessarily mean it's dead. In fact, it's often quite a good indication that it's still alive. Instead, poke it with a long stick and watch for a reaction. If you get one, attach your knife to a long pole and stab the beast in its neck muscles. If you think it's safe to approach, use a stout stick to bludgeon it on the head.

> Always aim to bleed and gut a large beast in situ before moving it. There are no prizes for carrying the heavy guts over any long distances for little reward!

MOVING LARGE ANIMALS

If you've caught yourself a deer, or an elk, or even a wild pig, moving it can be a challenge. They're heavy and awkward to handle, and if you try simply to drag them along the ground, you're going to expend crucial energy.

A better way to move them is to find a long, straight branch and tie the animal to it upside down. This can then be dragged by a single person, or one end taken by each of two people, as shown on page 183.

If your animal has large antlers, you should cut these off – or even remove the entire head – before trying to move it as they can get in the way.

FIELD DRESSING LARGE ANIMALS
Bleeding

We bleed large game for the same reason as small game (see page 175), but, because of its size, the process can be more difficult. If you can, suspend the animal from the branch of a tree. If that's not possible, you can make a frame from stout, straight lengths of wood. If the animal is very big, or if there aren't any trees nearby, you'll just have to deal with it on its side. I did this with a camel in the Sahara and it's totally do-able with a bit of patience.

Cut the throat or remove the head of the animal to bleed it, and remember to put a clean container underneath if you intend to drink or cook with the blood.

Skinning

As with small game, we start on the underside. If the animal is male, it's best to remove the testicles first. Being very gentle with your knife (you don't want to pierce the innards), make an incision lengthways between the legs just deep enough to slice into the skin. Cut towards and around the crotch, and then up the length of the animal to the neck. Put your knife to one side, get your hands beneath the skin and start to roll it away.

The better you are at butchering, the less you will use your knife. Work with your fists clenched and face down to 'knead' the skin away from the flesh in a pushing motion. On a fresh animal, the skin should separate fairly easily, although you may need to use your knife to slice some of the connecting tissue. To remove the skin from around the legs, you'll need to cut along them, and you'll need to make some cuts around the circumference of the legs just above the first joint in order to remove the skin fully.

> Being able to remove the skin from a large animal in one piece is a great survival skill: the hide can be used to provide much-needed warmth in cold environments.

Gutting

Once you've skinned your quarry, you must gut it. You can make a big slit along the length of the belly and simply scoop out the insides. Messy, but it would get the job done – the entrails come out in one big mass (although you may have to get your hands deep inside the carcass and make a cut where they're attached to the body). However, in a survival situation you might need to think a bit smarter. The insides of a large animal might look a bit disgusting, but they can still be a good food source if you know what you're doing.

If your kill is suspended, cut carefully – and not too deeply, to avoid slitting the internal organs – upwards towards the anus. You should find that the entrails spill out but still hang from the carcass, allowing you to inspect them. You can now start to cut away the bits of offal that you intend to use.

OFFAL AND OTHER BODY PARTS

The offal of large animals should be used as quickly as possible before it spoils. As with small animals, you should definitely help yourself to the liver (though husky and moose liver can contain toxic levels of Vitamin A), kidneys and heart. In a survival situation, the liver and heart can be eaten raw (many indigenous people do this), but if you can it's best to cook them. Some indigenous people, when they've butchered a large animal in the field, will place the liver directly on to hot coals to cook it through before eating it. A great source of nutrients and energy if you've gone without meat for a long time.

However, they're not the only parts of the animal's entrails that you can eat. The lungs are edible when boiled (make sure they are pink and free from any black or white spots – cut away any bits that look dodgy). You should also concentrate on the stomach and the intestines. They both have a role to play in keeping you fed and watered.

Stomach

Don't be put off by the contents of your kill's stomach. It has already been semi-digested, which means it's more easily digestible for you. So, in the most extreme situations, you can boil the contents of the stomach then eat it yourself to get some essential nutrition inside you.

The stomach itself is also edible. You've heard of tripe, right? To prepare it in the field, first scoop out the contents. Now wash the stomach really well in several changes of fresh water. Next boil it slowly. Not everyone's favourite meal, but when you're hungry . . .

It's also worth remembering that a stomach has been designed by nature to be a container. You can use it for that purpose too, by turning it into a water holder. Empty the stomach of its contents, then wash it out several times, until the water comes out clear. Now carefully turn the stomach inside out and, with the sharp edge of your knife, scrape at the inside lining so that it comes away without puncturing the stomach itself. Wash thoroughly. If possible, you should soak the stomach in warm water for a couple of hours so that it is thoroughly clean. You're now ready to use it as a water bottle. Don't fill it too full, and use a length of cordage to tie the opening tightly.

Intestines

The intestines of a large animal will look disgusting – long tubes of semi-digested food – but don't be put off. Chances are you've eaten something similar before, because they're often used to make sausage skins. And you can do the same thing in the field.

Intestine Sausages

First, squeeze out the contents of the intestines. Do this carefully, so you don't tear them. Now, carefully turn the intestines inside out and wash away any remaining undigested food. As with stomachs, you need to make sure that intestines are scrupulously clean before using them to make sausages. The best way to do this is to boil them, so get some water boiling over your fire and simmer the intestines for a good couple of hours.

They're now ready to fill. Cut away some meat and fat from your carcass and chop it up as small as you can. (Don't be shy with the fat – it'll make the sausage taste better and is a great source of energy.) You can season it with salt, pepper and spices from your trail mix, or with some chopped-up edible plants (wild garlic would be awesome). You can also mix in some blood, if it's fresh and uncontaminated (think black pudding).

Carefully stuff the sausage skins with your mixture, again taking care not to split them. When they're full, tie a knot in either end, or tie each end tightly with some cordage (see page 258). You can now either boil or fry your sausages. They can also be smoked to preserve your food supply for longer (see page 195).

Brain

If you can get the head of your animal open, the brain is a good food source. Just cook it like you would any other part of the beast – boil, grill or fry.

Tongue

Perfectly edible. Cut it out of the mouth and boil it whole. Peel the skin off before eating the flesh. Native Americans used to kill buffalo simply for the tongue: they removed it without having to field dress the animal, and would have a full meal on their hands. Nowadays, of course, we'd want to make sure we used as much of the animal as possible.

Bones

Once you've stripped your animal of its meat and offal, it still has something to offer you. The bones are full of gelatinous marrow and nutrition. Break them up as much as you can, get them in a pot, cover them with water and simmer for as long as you can. The resulting broth is life-giving. You can also use it as the base for wild soups and stews.

COOKING THE FLESH

Before you cook the meat from a large animal, you'll need to cut it into smaller, more manageable pieces. You can spend ages learning how to butcher an animal neatly and efficiently, but in a survival situation your butchery is likely to be a bit more rough and ready.

First, remove the hind legs. You do this by cutting round the ball-and-socket joint where the leg meets the body. You'll find bits of sinew around the joints, which you'll need to cut through. Then yank the legs back so that the socket breaks. You might need to get your knife into the joint a bit before the haunch comes away.

Now cut along either side of the spine and start removing the slabs of flesh on both sides of the body, using your knife gently to separate the muscle from the spine and ribs. Once these are away, you've got the bulk of the animal's meat, but you can break off the ribs, which will still have flesh clinging to them, then pick the remaining scraps from the carcass.

You'll now have two haunches, some thick slabs of meat, and some scraps and bones. You can spit-roast or boil the slabs and haunches. Smaller scraps, and any meat still clinging to the bones, are best boiled.

I also want to show you a really great way of cooking a haunch. It's been used by indigenous people for thousands of years.

First, dig a pit at least twice the size of the joint of meat you want to cook. Fill it with large flat stones, then light a fire above the pit. Let the fire burn for a good couple of hours – this will get the rocks really hot.

Carefully brush the embers away from the pit. Move some of the hot rocks out of the pit and arrange the remainder so you have a flat base and a surrounding wall of rocks. Place your meat directly on top of this. It will sizzle and start to shrink slightly. Cover with the remaining rocks.

You now need to insulate your makeshift oven. Place some long green (so they don't burn) branches across the top, about a hand's-width apart. Now cover with a thick layer of moss. Once the moss is in place, cover the whole thing with wet mud or sand.

You've now created an airtight, well-insulated oven. It will cook a haunch of deer to perfection in 2–3 hours. When the time is up, brush away the sand and carefully (everything will still be very hot) peel back the moss. Remove the sticks and lever away the rocks on top of the meat. It should smell fantastic, and the meat itself should be really tender.

There's nothing to stop you cooking a whole animal in this way. You'll just need a bigger pit and more cooking time. The result will be awesome, though. I have done this a few times with great excitement from those travelling with me. I think it's the suspense and smell of the slow-cooking, hidden, steaming, sizzling carcass in the ground!

Bear and Bacon Kebabs

If you've bagged yourself a bear (black ones make better eating than grizzly ones, but I'll say this again – be careful in bear country), you've got a whole load of meat to get through. Once the animal is skinned and gutted, cut the meat into cubes. If you've got it, wrap each cube in a rasher of bacon – or any other fatty meat – then thread them on to a wooden skewer and roast over a fire. Make sure the bear meat is very well cooked through, as it can harbour certain parasites that must be destroyed by cooking. (The same principle goes for cooking crocs – dangerous to catch, but they have a ton of amazing, rich, nutritious meat on them; see page 239.)

Campfire Stuffed Elk Heart

Take the heart from a freshly killed elk (or deer, or moose), trim it and give it a good wash. Slice off the top of the heart and then slit along its length to open it up. Clean out the inside, getting rid of any congealed blood. Now dice your stuffing – a mixture of apple, onion and garlic is good, with plenty of seasoning, but really you can use whatever you have to hand. Fold the heart up, replace the 'lid', then wrap it in tin foil and roast in the embers of your fire for about an hour.

BIRDS

Birds of prey – I'm talking about vultures, crows, seagulls, eagles, hawks – are the easiest to catch because they're so inquisitive. Unfortunately, they don't taste as good as game birds, which are harder to catch but a lot better for eating. All birds, however, are edible, and the process of preparing them is largely the same.

If you've managed to catch a bird, the method of preparing it for the pot is much the same as for small and large game. It needs killing, bleeding and gutting. The only difference is that you'll need to decide whether to pluck it or skin it. From a nutrition point of view, plucking is better because there's lots of good, energy-giving fat in the skin.

FIELD DRESSING BIRDS
Killing and bleeding

The best way to kill a bird is to twist and stretch its neck in one swift movement so that it breaks. Quick. Painless. Job done. Once it's dead, you can cut its throat and hang it upside down to bleed. Alternatively, you can kill it by cutting its throat and so combine the two jobs.

Plucking and skinning

It's easier to pluck a freshly killed bird while it's still warm. Pluck carefully and slowly, otherwise you risk splitting and tearing the skin. You'll probably find it easiest to start around the breast area then move round. It requires a bit of patience, but it's worth it.

You can make the plucking process easier by submerging the dead bird in almost boiling water for 30 seconds, letting the water drain off and the body steam-dry, then repeating twice more.

Skinning a bird is easier than plucking it. Slit the skin along the breast bone, then gently peel the skin (and feathers) away from the meat. When you reach the leg joints, bend them back until the joint snaps and cut the feet off with a knife. Cut off the head, then continue peeling off the skin until it comes away from the body. Use your knife to cut off the wing tips and the tail.

Gutting

Put the bird on its back and make a cut from just below the breast bone to the anal opening. Pull the cavity apart and remove the insides with a scoop of your hand. You can eat the heart and the kidneys.

Cooking

Older birds, and those that have been feeding on carrion, are best boiled – the old birds become more tender, and boiling prevents the risk of parasite infection from the meat-eaters. Otherwise, attach the bird to a spit in the same way you would a small animal. For a small game bird, where most of the meat is on the breast, you can just cut that away and cook it as you would any other piece of meat.

BIRDS' EGGS

Where there are birds, chances are there will be eggs. In many parts of the world it's illegal to touch them, and you certainly shouldn't interfere with nests thoughtlessly. But in a survival situation, they're a ready stash of protein, carbohydrate and fat. Pretty much all birds' eggs are edible. The best way to cook them is by boiling, but you can make a small hole in either end with your knife and roast them over a fire. Be warned: I've cracked open a bird's egg to find a semi-incubated chick inside. Not my favourite meal ever, but edible and a perfectly good source of food (once you get through the yuk factor of chomping through bone, feathers and blood!). In a survival situation you can't let anything go to waste.

Campfire Roast Duck

If you've managed to catch some kind of water fowl (see page 157 for how to do this), here's something to try. Pluck and gut the bird, then rub the skin all over with salt and pepper, plus any herbs you have to hand. Wrap it in tin foil and make sure the seal is tight, then place it over the embers of your fire for about 2 hours. Ducks and other water birds have a thick layer of fat to keep them warm. This will form loads of juices in which the bird will roast, but you'll want to get rid of the melted fat before you eat the bird. So, when it's nearly cooked, pierce the tin foil and the bird's skin with your knife and let the fat drain out. Now unwrap the bird and tuck in. It will be worthy of any feast day!

10

PRESERVING
YOUR KILL

If you've managed to trap, kill and butcher a decent amount of game, there's a good chance that you'll have more meat on your hands than you need in one sitting. In a survival situation, you can't let anything go to waste. In addition, you owe it to the animal you've just killed to make the most of the flesh with which it's provided you.

As we've already discussed, the offal should be eaten as quickly as possible before it spoils. There are, however, some good ways to preserve the flesh so that you can continue eating it days, or even months, after the kill. They are: freezing, drying, smoking and salting.

FREEZING MEAT

If you're in sub-zero temperatures and there's snow and ice on the ground, nature's already given you the perfect food-preservation device.

You must still skin and gut your game in the usual way. The quicker you get the flesh frozen, the longer it will last. So cut it into small pieces before packing it in snow or ice (this also has the advantage that you only need to defrost what you need to eat at any given time).

In winter, you'll probably find that your meat freezes solid out in the open, depending on the temperature. During the summer months, you should dig a hole in the snow and pack your meat into that.

When you want to cook your frozen meat, let it defrost slowly near an open fire before cooking through thoroughly in the usual ways. And if raw meat has defrosted, don't re-freeze it. Get it cooked, quick.

AIR-DRYING MEAT

You've probably heard of jerky – strips of dried meat that can last for ages. They're a brilliant food, whether you're in a survival situation or just on the trail. And knowing how to dry your own meat in the wild is a great way of planning ahead.

You can air-dry small animals (anything up to the size of a squirrel) whole. Make sure the animal is bled, gutted and skinned. Suspend it off the ground in a sunny place with good airflow. The meat will start to dehydrate. Once it feels dry, break it up with a rock to expose the bones and the marrow – this will not be as dry as the exterior and will go off if you leave it that way. Leave the meat to dry in the sun for a second time.

If you want to dry lean flesh, you need to cut it into strips, about 3cm wide and 0.5cm thick. Make sure you trim it of as much fat as possible – fat doesn't dry well and will spoil the meat. Hang these ribbons of meat in direct sunlight – a makeshift drying rack constructed from two branched sticks with a pole stretched between them will help you do this. Make sure that none of the strips of meat are touching each other.

Alternatively, thread your strips of meat on to a length of wire or a thin branch and suspend that above the ground. Again, make sure none of the strips are touching.

Dry the meat in the sun until it has a really crisp texture – if the strips are hanging over a drying rack, you want them to snap at the bend if you pull them. It'll take up to 24 hours, but the meat will keep indefinitely as long as it's kept dry.

Air-dried or smoked meat retains all the nutrients of the original, undried meat, but because you've got rid of so much water, it can weigh less than half – useful if you're on the move with a heavy pack.

SMOKING MEAT

Air-drying meat relies on sunny, airy conditions. But in the absence of these, you can also preserve meat using smoke. It dries the meat out and creates a smoky layer on the outside that keeps the bugs away.

The best way to do this is to dig a small pit and light a fire in the bottom. When the fire has burned down to glowing embers, add some twigs of green hardwood (soak it if it seems very dry). Unless there's nothing else about, avoid pine or other resinous trees. The smoke's not harmful, but it will taint the taste of the meat.

Cut and suspend your strips of meat as for air-drying, then hang them a couple of feet above the smoking coals. The fire shouldn't be too hot – you're not trying to cook the meat, just dry it out with the smoke.

Keep the fire smoking for as long as it takes to dry the meat out: you want it to crack when you bend it. This can take up to 24 hours.

You can grind air-dried or smoked meat down into a powder called pemmican. A brilliant survival food – eat the powder raw or add it to any of the soups or stews in this book to boost the flavour and the nutritional value.

You can smoke fish in exactly the same way that you smoke meat. Just make sure it's freshly caught and that you're confident before you start that the flesh is safe to eat. See pages 98–103 for the low-down on fish.

Many spices have anti-microbial properties. That means that they inhibit the growth of bacteria in food. Spices with this quality include garlic, chilli, black pepper, ginger and cinnamon. If you have dried spices in your trail box, rub them into your meat before drying it. This will not only make it taste better, it will also slow down its deterioration.

SALTING MEAT

One of the oldest ways of preserving meat (or fish) is to use salt, either by rubbing dry salt into the flesh, or by soaking the flesh in a brine solution. The salt draws out the moisture and the dried flesh will keep for a long time. This is not particularly practical in the field, because it can take a long time to salt the flesh fully. But if you salt the flesh *before* drying or smoking it, it can speed up the process and also make it taste better.

If you're near the sea, collect some sea water and boil it down until crystals almost start to form. This concentrates the salt solution and kills any bugs in the water. Let it cool, then soak your strips of meat in the solution for half an hour or so before proceeding to dry or smoke them as above. You'll probably want to wash off any excess salt before eating the finished product, or even soak it in fresh water if it's very salty.

PRESERVING MEAT IN FAT

You can preserve meat for shorter periods of time using fat. This can be difficult to do in the field, but it's a good trick to have up your sleeve because it doesn't take as long as air-drying, smoking or salting. The meat needs to be cooked first to kill any bacteria. You then need to render down any fat from the carcass and place the meat into it so that it is completely covered. As it all cools down, the fat will congeal and protect the meat from bacteria for a few days. It's best to reheat meat that has been preserved this way before you eat it. Just scrape off as much of the fat as you can and heat it through in a dry pan (or on a hot, flat stone placed close to the fire). Make sure it's thoroughly hot before you eat.

PART THREE

THE WHOLE HOG

11

INSECTS (AND OTHER CREEPY CRAWLIES)

D id you know that two-thirds of all living creatures are insects? They're the most abundant complex life form on the planet. By a long way. And that means there is a lot of food out there — if you know where to look and can get over the whole thing of eating bugs.

As the population of our planet increases, so will the demand for food. Our traditional sources of meat and fish won't be enough to feed us all. There are many experts who believe that we'll be obliged to make insects a part of our diet in the next fifty years or so: lots of them are edible, and they're everywhere.

Think about what that means to you in a survival situation. You might be finding it difficult to get your hands on meat or fish, but there are probably perfectly good food sources nearby, if you only know where to look.

Perhaps you think that eating insects sounds disgusting? Well, most of the world's population eats them on a regular basis. Insects and other bugs form an important part of the diets of many cultures around the world, from Asia to the Amazon, and for many indigenous people a plague of locusts is like the skies raining food.

And high-nutritional-value food too: most insects are an excellent source of protein, carbohydrate and fat. Take a small grasshopper: 100g of grasshoppers provides you with about 20g of protein, 6g of fat and 4g of carbohydrate. Compare that to 100g of beef, which will give you about 27g of protein, 5g of fat and hardly any carbohydrate. In a survival situation, that little grasshopper is a real all-round super-food.

Most insects are edible, but you should avoid any that have very brightly coloured markings. That's nature's warning to stay away, and although there are some edible types that are brightly coloured, it's worth avoiding them just to be safe. Insects are also mostly edible raw – though, as with all raw foods, there's a risk that they might carry parasite infections and the only way to get rid of these is by the application of heat. So cook your bugs if you can – you'll find some ideas for how to do this in the pages that follow.

The best place to find insects during the day is in cool, shady places. They will burrow into the bark of trees, or congregate in dark, damp holes. If you're lucky, you'll find beetle larvae. These can grow up to 15cm long and are particularly nutritious – if you can swallow them down without gagging!

This chapter is not only about eating insects: it's about other creepy crawlies too. Technically, insects are part of a group of creatures called arthropods, which also includes arachnids (that's spiders and scorpions to you and me) and crustaceans. I'm also going to throw in the occasional gastropod (snails and slugs), which can be an excellent source of food.

CRICKETS

Where to find them: all environments except Antarctica.

The Latin name for cricket is *Gryllus* (I'm not kidding, I promise!). And crickets are one of the most frequently eaten insects known to man. In many parts of the world they're thought to bring good luck (which they certainly do if you're hungry and come across them in the field). They're a common snack food in Thailand

and Cambodia, and of all the insects that you might find yourself eating in a survival situation, these are one of the most palatable. They really don't taste too bad – although be prepared for a slightly fishy flavour that can catch you unawares!

Crickets are pretty easy to catch. You'll find them mostly in warm, humid areas where there is a decent amount of plant material to eat. If you're in a place where they are particularly abundant, you can catch them by swiping a tightly woven net across tall grass or foliage – try doing this in the early morning, when the cooler temperature makes the crickets very sluggish. Alternatively, you can use a bottle trap (see page 154). A third method is to use a plastic bottle. Remove the cap, cut off the top third, then put this top section upside down into the base. Sprinkle some sugar into the trap so that it lightly covers the base. Leave it in an area where you know there are crickets. The insects will be attracted by the sugar and able to enter the bottle, but not escape it. (You may find a few other delicacies get inside as well!)

You can eat the whole cricket, but you might find it more palatable to remove the wings, legs and antennae. Crickets can then be eaten raw (though you should try to cook them if you can), boiled, roasted or fried. My favourite method is to get a pile of them and simply fry them up, nice and crispy, with a drizzle of honey!

You can also use crickets to make a tasty trail snack. Here's how.

Bear's *Gryllus*

Drop your crickets into some boiling water and cook them for a couple of minutes. Remove and let them dry out. Lay them out on a piece of tin foil and place them over the embers of your fire. Alternatively, place them over the fire in a Dutch oven (see page 35) or any other cooking vessel you have to hand. Cook for an hour or so. You want them to be completely dried through – you should be able to crush them easily between your fingers. Once they're dried, let them cool, then gently roll each cricket between your palms to remove the legs and antennae. Sprinkle with salt, or whatever seasoning you have in your spice mix. These salty, crunchy crickets make a great, nutritious snack in the field.

Only male crickets make the distinctive chirping sound. The warmer the weather, the faster they chirp. When they get very cold, they don't have the energy to chirp – but that doesn't mean they're not around!

Although crickets have external wings, most types can't fly. But look out! Some types of cricket can bite.

ANTS

Where to find them: all environments except Antarctica.

Ants are awesome food. There are tens of thousands of different kinds, and they're one of the most abundant life forms on the planet – the total biomass of ants is equivalent to the total biomass of humans. That's a lot of ants. They might look pretty harmless when they're crawling around on the pavement, but in the tropics vast armies of ants can kill much bigger creatures as they swarm through the jungle – and they can carry fifty times their own body weight. In short, never underestimate the humble ant!

Most ants are edible, though you should cook them first to destroy a toxin many of them harbour. (There are certain stinging ants you should avoid, like fire ants, whose stings can make you very ill. The bigger the ant, the worse the sting.) Lots of ants have a vinegary, citrusy taste because of the formic acid they contain – they explode in your mouth when you eat them and release the acid! I actually love the taste of this.

In Australia, honeypot ants are particularly prized by aboriginal peoples. The ants eat so much that they swell to the size of small grapes and are filled with a sweet, nectar-like liquid. They can be dug up from the ground and eaten raw. In South America, leafcutter ants are toasted and eaten like peanuts. And in the rainforest, tiny lemon ants are a delicacy.

Weaver ants have a particularly bad sting – they spray their formic acid into their bite, which really packs a punch. But their larvae contain more protein than the same quantity of beef, so if you can put up with the bites they're a good survival food waiting to be harvested. If you come across an agave plant, it's worth inspecting the roots. There's a good chance you'll find ant eggs here. In some parts of the world they're known as 'insect caviar'. The eggs, once collected, can be boiled or fried and have a good buttery taste. Yum!

You can collect ants by leaving a scrap of food out for them. If you've ever been on a picnic, you'll know how quickly they'll arrive. You'll need a lot of ants to make a decent meal, but you can use them to bring some flavour to other food you might be eating in the wild – fried ants are pretty good sprinkled over wild greens as a condiment.

In South America, queen 'big-butt' ants fly in swarms of thousands and can be caught in large nets ready to be toasted and eaten.

BEETLES

Where to find them: all environments except Antarctica.

Of all the edible insects, beetles are one of the most widely eaten – perhaps because 40 per cent of all insects are beetles. The scientist J. B. S. Haldane once said that God must have an 'inordinate fondness for beetles'. But I like them because they can be the ultimate survival food.

One of the most popular types is the dung beetle, which you'll commonly find feeding underneath cow pats. That might sound a bit disgusting, and although I'd normally steer clear of insects that are feeding on faeces, really cow pats are just semi-digested grass and, when you're hungry in the wild, you can't be too sensitive about this stuff. Dung beetles are edible, as are scarab beetles, rhino beetles, longhorn beetles and June bugs.

The fully grown specimens of these beetles are great survival foods. Remove the head and wings, then wash them well in fresh water before boiling or frying.

But you can also eat (and I often have) beetle larvae, which can be absolutely massive – a meal all on their own. These pale, sausage-like grubs are best found hiding behind peeling tree bark or in the cool, dark crevices of tree stumps. They look a bit like caterpillars, but are shaped more like a 'C'. Beetle grubs are an amazing survival food because they're so high in both protein and fat. Sometimes in the tropics you'll come across bamboo shoots that are completely infested: a real treasure trove for the hungry survivor! One of the most commonly eaten beetle larvae is the mealworm, which is produced commercially and turned into a highly nutritious flour.

Both beetles and their larvae are a rich source of protein and trace minerals. Eat them raw if you have to, or fry them up to make them taste great, mixed with a little maple syrup!

BEES

Where to find them: all environments except Antarctica.

Coming across a swarm of bees is a good sign for three reasons. First, their presence almost always indicates that there is water nearby. Second, honey bees mean honey. And third, you can eat them. Bees like to nest in dry, dark places. Think abandoned rodent holes, dark corners of dilapidated buildings, caves . . .

Of course, bees present one big problem. Or rather, one little problem: their sting. A single sting from a desert bee once made my entire face swell up until I was barely recognizable, and if you're particularly sensitive to bee toxins, there's a real risk of anaphylaxis and death. If you do come across a bee swarm, however, there are several smart ways of turning it into a food source if you judge that risk worth taking.

The best time to gather bees is towards nightfall. Worker bees travel miles from their nests, but always return there at night (they have an incredible sense of direction).

With bees, smoke is your best friend. If you light a fire (use lots of green vegetation to make it smoky) and blow the smoke in their direction, it will sedate them and make them less likely to dive-bomb and sting you. (This might sound counter-intuitive, but they have evolved to start feeding when they encounter smoke, so that they have enough energy to leave the hive if the queen decides that they must escape. And a full bee is a sleepy bee.) So, if you create a lot of smoke and keep it going for a couple of hours, you might find that you can drive honey bees away from their nest so that all you're left with is the honeycomb and, if you're lucky, the bee larvae. These larvae grow in the cells of the honeycomb and they're an amazing food source – sweet-tasting, made up of almost 50 per cent amino acids and stuffed full of minerals and vitamins, especially (no pun intended!) B vitamins. Then, of course, there's the honey itself: a great instant fix of carbohydrate that will last, literally, for ever. (Honey doesn't spoil or deteriorate – which is why the Egyptians used honey in the mummification process.)

To catch the actual bees (which are also incredibly nutritious), your best bet is to find a swarm that has made its nest in a hole. You can kill them by filling the hole with smoke and then covering it. This will give you access to all the food inside: the honey, the larvae and also the bees themselves.

If you're hunting bees, get yourself covered up as much as possible. Tuck your shirt into your trousers and your trouser legs into your socks, if you're wearing them. Cover as much of your face and head as possible with a T-shirt or another piece of cloth. If you do get bees on your skin – and you probably will – don't brush them off because that's when they sting. Leave them to fly away naturally.

Once you've harvested them, you should remove their wings, legs and – crucially – their stings before eating. (You'll find the stinger at the rear of the bee's abdomen. Depending on what type of bee it is, it's sometimes barbed and will be attached to a small venom sac. The best way to remove the stinger is by using a pair of tweezers (you might have these in your First Aid kit or penknife) – hold the bee's abdomen with the fingers of one hand and give it a little squeeze to make the stinger protrude.) Be careful.

Bees can be boiled or fried. Roasted bees make a great snack.

CENTIPEDES (and not millipedes)

Where to find them: all environments except Antarctica.

Don't fall into the trap of lumping centipedes and millipedes together. They're very different creatures.

Centipedes don't have a hundred legs and millipedes don't have a thousand, but you can tell the difference by counting the legs on each section: centipedes have one pair of legs on each body segment, whereas millipedes have two.

The differences don't stop there. Millipedes are generally not edible: there are species that contain hydrocyanic acid, which in high enough quantities can kill. You're unlikely to die from eating a single millipede, but it can burn badly and make you feel quite unwell, so you should avoid them.

Centipedes are a slightly different matter, although they still have to be handled with care. I've come across giant centipedes that are highly venomous, with big front claws which are able to deliver a dose of poison that will make you feel as if you're being stabbed with a red-hot poker. (Not what you want scuttling up your leg and then biting you!) I abstained from eating that little beauty, but there are certain centipedes, like the giant red-headed centipede *Scolopendra heros*, that are prized as a food in certain cultures. You need to make sure you remove the head and the first set of legs, which are in fact toxic pincers – they use them to subdue their prey. The remainder of the centipede can be cooked and eaten.

COCKROACHES

Where to find them: all environments except Antarctica.

I wouldn't advise eating any cockroaches you might find around your house, but out in the wild they can, despite their reputation, be a clean-living bug. Wash them first, then either boil them or fry them.

If you have some sugar in your pack, sprinkle this on to the frying pan so that it caramelizes over the roaches. If you can find younger specimens, they tend to be better to eat because their outer shell isn't quite so hard.

> You can despatch most insects by removing their heads. Not so with cockroaches – they can live for a week without it! Tough, eh?

DRAGONFLIES

Where to find them: all over the world except Antarctica.

Dragonflies are fast – they can fly up to 35 miles per hour – but if you catch one you've got yourself a meal. You'll normally find them near water. If they're very abundant, you might even be able to catch them with a net. In Indonesia there is a traditional method of catching dragonflies that involves taking a bamboo pole and tying a long strip of palm to one end. The palm is then covered with the sticky sap of the jackfruit tree. You flick the palm towards the dragonfly; if it touches it, the insect sticks and you've got dinner. Alternatively, you can wait until nightfall, then take a torch to the water's edge and examine the underside of leaves. This is where the dragonflies rest up at night, when they're much slower – and you should be able simply to pick them off the leaves.

You'll need to remove the dragonfly's wings before eating it. What's left is good fried, or you can skewer several on to a straight twig and toast them over the fire.

Dragonflies spend most of their life cycle as nymphs, which live in the water and are also edible.

If you're trying to catch dragonflies, wait for the sun to go behind a cloud. They need warmth to fly – when the sun disappears, they tend to settle on the ground.

Damselflies look similar to dragonflies. You can tell the difference because unlike dragonflies, damselflies can fold up their wings. Both are edible, though.

EARTHWORMS

Where to find them: all environments except the Arctic and Antarctica.

Worms are full of protein and very abundant wherever the soil is rich and moist. They'll come up to the surface after a rain shower, or you can dig for them. That's one of the reasons they're such a great survival food. You might not know how – or have the tools – to hunt or fish or trap animals. But you sure know how to dig (and you have the tools to do it – they're at the end of your arms!).

It's possible for there to be more than a million worms in one acre of land. And your humble earthworm is more than 80 per cent protein and very high in iron (in some parts of South America they are eaten by pregnant women for this reason).

Worms eat dirt, and any worm you find will be full of the stuff. There are a few ways to deal with this.

Probably the best is to starve them for 24 hours, during which time they'll purge themselves. But if you don't have 24 hours before you need to eat, you can drop them into clean water for 10 minutes. This will cause the worms to clean themselves out.

Alternatively, you can gently squeeze the contents of the worm's body out from one end, like toothpaste from a tube.

Earthworms *can* be eaten raw (just make sure you wash them first), though as with all creepy crawlies it's probably best to cook them first – 10 minutes in boiling water is fine. After that, you can add them to wild stews, fry them, or just eat them whole.

My personal favourite is the worm omelette. A couple of pigeon eggs and a handful of worms and you have a great dinner. The cool bit is that it looks so disgusting, but actually doesn't taste too bad!

You can smoke earthworms just like you smoke meat – see page 195.

Dried earthworms can be ground down into a highly nutritious powder – a great thing to add to your wild food survival recipes to boost the protein and iron content.

Earthworm Jerky

Take several large earthworms and clean them by gently squeezing out the contents. Boil for 10 minutes, then skewer them on to a thin twig. Find yourself a flat stone and heat it in a fire. Carefully remove the stone but keep it near the fire's embers and lay the skewered worms on it. Keep them there for 10 or 20 minutes, turning frequently so they don't stick to the stone, and making sure they don't burn. You want them to be nice and stiff. Remove from the twig and eat.

Earthworm jerky will last for ages – Native Americans used to dry earthworms in the spring for use during the lean winter months.

FLIES AND MAGGOTS

Where to find them: all environments except Antarctica.

Most flies are edible, but you should avoid any that have been feeding on anything disgusting, like animal faeces. There are stories of people surviving for months in the harsh Australian outback eating nothing but the flies they could catch, and in Africa vast quantities of them are mushed up and turned into a great survival food called kunga cake.

And don't forget the maggots – an awesome food source. The only trouble with maggots is that they're not too fussy about what they eat, so you're most likely to find them on rotting debris or decomposing flesh. Normally it's an indication that the flesh is none too fresh, so you shouldn't eat it. The maggots can be eaten, though you should purge them first by starving them for 24 hours so that they expel any of the rotten flesh that they've been consuming. Once they've been purged, you can boil or fry them before eating.

Avoid any maggots that you find on animal faeces.

GRASSHOPPERS AND LOCUSTS

Where to find them: all environments except the Antarctic.

Grasshoppers and locusts are essentially the same thing. The Bible tells us that John the Baptist lived in the desert on a diet of locusts and honey. He must have been something of a survival expert as well as a visionary (in fact he is one of my heroes – the ultimate wild man on a mission!), because locusts are a great source of food. They're widely eaten across South America, Africa, the Middle East and Asia.

The best time to gather grasshoppers is early in the morning when they move slowly because of the relative cold; you can simply pick them off blades of grass. However, when food becomes scarce they congregate in big swarms which can fly massive distances looking for vegetation, and it's when they're swarming that they become known as locusts. Locust swarms can be pretty scary, blackening the sky for miles around, but as they pass through, you normally end up with a good number of locusts on the ground which you can harvest and eat.

Grasshoppers and locusts are edible raw in extremity, but you should aim to boil or fry them if you can, or even skewer the larger specimens and roast them over your fire. You'll probably want to pull off the legs and wings first to make them a bit easier to get down your throat.

SCORPIONS

Where to find them: all continents except Antarctica and high-latitude tundra.

There's an old Cantonese saying: the Chinese will eat anything that flies, except an aeroplane; anything with four legs, except a table; and anything that swims, except a submarine. It's pretty much true: walk through a Beijing food market and you'll find anything from dog-brain soup to stewed goat lung.

A French historian called Jean-Baptiste du Halde wrote in 1736 about a Chinese banquet where guests ate stag penises, bear paws, cats and rats. So, to a hungry Chinese person, scorpions are decidedly tame – although quite a delicacy. You can buy them frozen in bags a bit like we buy frozen prawns. In fact, they even taste a little like large prawns, and you eat them with the shell still on. The desert hairy scorpion (*Hadrurus arizonensis*) is one of the tastiest, but they're all edible.

Scorpions can be eaten raw or cooked. If they're raw, you'll need to cut off the poisonous stinger with a knife. (Be warned – raw, they taste pretty terrible, full of pungent goo that has one hell of an aftertaste.)

By the way, if you're cooking them, the heat will break down the proteins in the venom, making it harmless – but you might still want to cut the stinger off anyway. Once you cook scorpions up, though, they start to taste a lot better – crunchy and nutritious.

So all round, scorpions are a very good, high-protein survival food. If you find a live scorpion in the wild, pin it down with a stick, cut off the stinger with a sharp knife (be careful, because it will try to strike repeatedly when it knows it's in danger). Then cut off the pincers to stop it biting you and just pop the little critter into your mouth. It could be the difference between starvation and making it through another day. I think I must have eaten hundreds of them in total!

If you have more time – and more scorpions – you can make a better meal of them. If you're in a region where scorpions are rife, hunt for them in dark places – under big rocks, or hidden in logs (they are largely nocturnal). Be very careful, though – scorpion stings are nasty and sometimes even fatal. (The rule of thumb is that the smaller the scorpion, the more deadly the venom.) Scoop them up with a net.

If you must hold them for any reason, do so by the tail, either side of the stinger, so that *you're* controlling the venomous bit, not the scorpion. And if it pinches you, do *not* let go of the stinger, as it will then be free to sting you!

It's traditional in some cultures to soak the scorpions in milk for half an hour before cooking them, to help with the taste. If you don't have any milk to hand, you can fry them just as you'd fry raw prawns, using whatever seasoning takes your fancy. Alternatively, you can skewer them on a long, thin stick like a kebab, then grill them over your campfire. Cook them really well and they taste a bit like shrimp (if you keep your eyes closed).

TARANTULAS

Where to find them: widespread throughout the southern hemisphere, except Antarctica.

Tarantulas get a bad rap. It's true that some of them can give you a very nasty bite, but they're not generally harmful to humans and some are not venomous at all. But they are edible.

Tarantulas are particularly popular in Cambodia. Cambodians started eating them during the brutal Khmer Rouge regime when food was in very short supply. They became a true survival food, and if you come across them in your own survival situation, you can consider yourself lucky.

The best tarantulas for cooking are the Thai Zebra variety (*Haplopelma albostriatum*). These are very common in Cambodia,

Thailand and Malaysia, where they are a popular fast food (and one that children love to hunt for). You can go to a Cambodian market and buy live, de-fanged tarantulas by the dozen from huge wicker baskets, ready to take home and cook.

Or you can catch your own. It's not too hard. The spiders live in little round burrows, the openings of which they cover with a fine, silky web. To catch them, gently tickle the web to imitate a trapped bug. When the spider climbs out on to its web to see what's for lunch, slide a shovel behind it to cover the opening and stop it retreating back into the burrow.

You can pick up a tarantula by grabbing its back with two fingers, just in front of the abdomen – this avoids the fangs. Worth doing, because a tarantula's bite can be nasty – like a bee sting, only worse. Avoid doing this with your bare hands: wear gloves to avoid the tarantula's urticating hairs (see below). Alternatively you can just pin them down with a stick and despatch them with your knife.

Once you've caught your tarantulas, the best way to kill them is to drop them in a bowl of water until they drown. This will help clean them as well. Or just skewer them with a sharp stick to kill them. You then need to locate and remove the fangs, and singe off any hairs. Once you've done that, they're ready to cook. It's very common to stir-fry the spiders in plenty of hot oil in a large wok. If you're cooking in the wild, though, it's probably easier to use the method overleaf.

A tarantula's bite isn't the only thing you need to be careful of. They have little needle-like hairs on their abdomen, which they scrape off and throw when they feel threatened. These hairs can cause a nasty rash and a bad allergic reaction. They are called 'urticating hairs' and if any get lodged in your throat as you try to eat a tarantula, your throat can swell up and suffocate you. So remember: the urticating hairs 'urt, so always burn them off over a flame before cooking.

Spicy Fried Tarantulas

After burning off the hairs, bring some water to the boil, then boil your tarantulas for 2 minutes. Remove and allow to dry while you heat up a little oil in your frying pan. Sprinkle the tarantulas with spices from your trail spice box – a good mixture is salt, sugar, pepper, garlic and chilli. Fry the seasoned tarantulas for 2–3 minutes, until the skin is nice and crispy and the meat on the legs feels solid. Eat hot – they will have a smoky, spicy taste.

TERMITES

Where to find them: tropical, subtropical and temperate regions.

Termites have a bad name in built-up areas because of their tendency to consume and destroy wooden structures. But for the survivor, termites are awesome. They've been a major food source for indigenous peoples of South America, Africa and Australia for thousands of years. Hardly surprising – pound for pound they're more nutritious than vegetables and have a higher protein and fat content than beef or fish (100g of beef will give you about 300 calories, 100g of fish about 80 and 100g of termite about 560).

Another reason they're such a great survival food is that they're so easy to harvest: they offer a vast amount of energy for hardly any expenditure. In tropical regions they live in huge, impressive 'termite mounds', which can each house millions of the insects. Seeing one of those in the wild is like seeing a restaurant, where the food is free and, for the survivor, incredibly nutritious.

There are winged termites and worker termites. The winged ones are larger and more prized by termite aficionados, but the worker insects are easier to catch. If you come across a termite mound, you can simply dig your knife into it and watch them pour out. You'll be able to pick them off one by one. Alternatively, if you poke a wet stick into the termite mound, you'll find it covered with the little beauties when you pull it out.

We'll ignore the fact that termites indulge in what scientists call trophallaxis (which means they eat each other's faeces). They are in fact very clean creatures, and groom each other to keep parasites at bay. Eaten raw, they taste a little bit zingy. Cooked, they taste kind of like hazelnut.

Fried Termites

Because termites have such a high fat content, you can put them in a dry pan and fry them in their own fat. If you have winged termites, you'll want to remove the wings before adding them to the pan, then fry them gently over your fire until they're brown and crispy.

Termites don't only live in termite mounds. In temperate regions they can be found – along with many other types of bugs – in tree stumps and rotting wood.

WASPS

Where to find them: all environments except polar regions.

Like bees, wasps are edible (so they are good for something – although you do not, of course, get the added advantage of the honey). Wasps are a lot more aggressive than bees, so you need to be even more careful when you're approaching a nest (though it's only the females that have stingers, not the males). But they can be caught, killed and eaten in the same way – make sure you remove the stingers first.

As with bees, wasp larvae are perhaps more commonly eaten than the adult wasp – use smoke in the same way as for bees to drive away the adults, then help yourself to the babies. They are best fried, with perhaps a little honey if you have it, to give them some extra sweetness.

Hornets are a type of large wasp. They and their larvae are extremely edible and you might be tempted to go after them because their size means a good meal. Only do that if you're really desperate: hornets are one of the most aggressive insects out there, and their stings are *really* nasty.

A dead wasp emits a pheromone that attracts other wasps. Get it eaten if you don't want to be inundated.

CICADAS

Where to find them:
temperate and tropical regions.

Cicadas are amazing. Some species spend up to seventeen years buried in the ground, without even poking their noses up into the sunlight. (Even the more short-lived types spend a couple of years underground.) Then, all at once, after nearly two decades of this weird hibernation, all the nymphs in a particular location will climb out of the ground in huge swarms. These young cicadas have no wings, so when they arrive they can be easily harvested for food. I've caught and eaten cicadas many times in Africa, and they're used as a source of food all over the world – they're a particular delicacy in Mexico.

In Africa, I was once shown how to beat them from their roosting places in the tree tops. But you can use your ears to find them at night – the sound of a cicada is totally distinctive. With a torch, look at tree trunks. When they emerge from being buried in the ground they crawl up the nearest tree and you can just pick them off.

Here's how I cook them – very simple, and you could use this method for most insects.

Sand-cooked Cicadas

Pick the legs off your freshly caught cicadas to stop them running away. Light a fire on a patch of dry sand. When it's been burning for a while and the sand is good and hot, carefully brush it away. Lay your cicadas on the hot sand. They'll cook through in a few minutes. Once the cicada is cooked, you can tug out its insides from the anal opening – they'll come out in one gloopy piece. Then just pop this piece in your mouth and eat. Easy!

GIANT WATER BUGS

Where to find them:
worldwide except Antarctica.

These fellas are also known as 'toe-biters', because that's just what they can do. They have a nasty little bite, so you should be careful when handling them. They can also be huge – some of them grow up to 12.5cm long. That said, they're pretty easy to catch. Giant water bugs are abundant in freshwater lakes and streams all over the world, but especially in the Americas, northern Australia and Asia. At night they are, like many insects, attracted to the light, so if you have a torch or a lantern, place it by the water's edge and you'll find that they come to you – scoop them up with a small net.

The bugs can be boiled, fried or roasted, then eaten whole – though lots of indigenous people prefer just to scoop out the meat from the inside.

SNAILS

Where to find them: worldwide, except Antarctica. They prefer cool, damp places but are sometimes found in desert climates.

Many land snails are edible, but some are very poisonous. Avoid any that have very brightly coloured shells. In particular, steer clear of the Conus – a brightly coloured sea snail with a vicious sting that can kill a human.

You can, however, eat pretty much any snail found in the UK, where I live. Perhaps the most delicious is the common garden snail, *Helix aspersa*, which is also found widely across the world. They come out to play when it rains, so if you go hunting for them after a rainfall, you're in for a feast.

However, unless there's no other option, you should avoid eating them immediately. Snails need to be purged of all the nasty stuff in their digestive system. (They can carry all sorts of parasites, including a particularly nasty one that they get from rat faeces.) The best way to clean them out is to put them in a container that has holes large enough to provide ventilation, but not so big that your dinner can crawl away. An ice-cream tub with holes punctured in the top of the lid with a skewer does the trick.

Put your snails in the tub and add a few lettuce leaves. Let them eat this for five days, then remove the lettuce leaves and starve the snails for another two days. Now they're ready to cook.

I should tell you that some snail-gatherers will eat their harvest without purging them first, and in a survival situation you might find yourself having to do this. If that's the case, you should make sure you cook them very well first to kill any parasites. And you should definitely avoid them if they've been feeding on poisonous mushrooms (although slugs are much more likely to have eaten these than snails – see page 224).

Bring some water to the boil, then chuck in your snails. Let them simmer for 10 minutes, then drain them and leave them to cool. You can now remove the snails from their shells with the sharp end of a needle, a toothpick or with the tip of your knife.

They might be a bit slimy. If this worries you, toss them in a little salt, leave them for 5 minutes, then rinse off the salt. It should take the slime away with it.

You can either eat the snails like this, or you can fry them up in oil or butter with plenty of garlic – or, even better, wild garlic leaves that you've foraged yourself (see page 43). That way you avoid the slime!

Snails hibernate in the winter, so if you want them you'll have to go hunting underneath stones, logs or anything else that provides them with shelter. It's worth doing – hibernating snails don't consume anything, so you don't have to purge them before eating. But you must still cook them.

SLUGS

Where to find them: Europe, Australia and
North America.

You can just think of slugs as snails without
their shells. Like snails, they need to be
purged and cooked before you eat them.
They can contain many dodgy parasites, but
they also sometimes feed on toxic mushrooms.
If you've read Chapter 3 on fungi, you'll know that you don't want to
get any of that in your system.

If you don't have time to purge them, you can split them open
and rip out the guts. These are the bits that will contain any parasites,
so discard. Now boil the meat for a few minutes until it's properly
cooked through. A bit slimy, and certainly not the most delicious
thing I've ever eaten – but good protein and energy if you need it. I
also tend to drink the water in which they've been boiled, so as not to
lose out on any precious nutrients.

WOODLICE

Where to find them:
all environments except the
Arctic and Antarctica.

Woodlice might look like insects, but in
fact they're a form of crustacean, and indeed
they taste a bit like prawns. Unlike their
seafaring relatives, they live on land – although
they do still prefer moist areas, so the best place to look for them is in
damp, dark places such as in shady old tree stumps or under stones or
fallen branches. They're easy to collect, and can be quite abundant
once you find their hiding places.

The best type of woodlouse for eating comes from the family *Armadillidiidae*. These are also known as pill bugs or roly-polies, because they have a tendency to roll up into a little ball when they're threatened. To cook them, simply drop them into boiling water, or fry them over your fire. If you do this, you'll find they'll 'pop' as they cook – I like to think of woodlice as nature's popcorn. When they stop popping, they're good to eat. You'll have a tasty, crunchy little ball of nutrition that's 90 per cent protein.

12

AMPHIBIANS AND REPTILES

As with insects, many people have a bit of a mental block when it comes to eating amphibians and reptiles. And as with insects, there's no reason for this to be the case – although it is true that the pursuit of snakes, crocs and alligators can be a lot more dangerous than the pursuit of bugs and creepy crawlies. A lot of the information in this chapter really is for use in a survival situation only. You shouldn't go after puff adders or saltwater crocs lightly, but if there's nothing else to eat, knowing how to deal with them could just save your life . . .

AMPHIBIANS

When we say amphibians, we mean, largely speaking, frogs, toads and salamanders. There's another group of amphibians called caecilians – wormy, snaky things that bury themselves in wet mud, but they're not good to eat, so we won't be worrying about those.

In wet, marshy areas, amphibians can be a valuable food source. But you've got to know what you're doing, because you can make yourself pretty sick eating them if you don't take a few precautions.

FROGS AND TOADS

Most frogs are edible. Most toads are toxic. So you need to know how to tell the difference.

First the confusing bit: all toads are actually frogs, but not all frogs are toads. Toads are just a sub-classification of frogs. They're also generally found in the same regions. Frogs live on every continent except Antarctica. Toads are indigenous to everywhere except the polar regions and Australasia – although some types have been introduced into Australia.

However, they do, in general, have some distinct differences:

1. Toads have dry, warty skin, whereas frog skin tends to be smooth and slimy. This is because of where they live: frogs like moist environments (they're always found near water), whereas toads prefer to keep dry (they can be found near water, but are also found far from it).

2. Frogs mostly have webbed hind feet (for swimming) and strong, long hind legs (for hopping). Toads have shorter hind legs (they walk rather than hop), and their feet tend not to be webbed.

3. Frogs live on the ground *or* in trees. Toads only live on the ground.

4. Toads are chubbier and stumpier than frogs. They don't move around quite as much or as quickly as frogs.

So now you've worked out if your amphibian is a frog or a toad. If it's a toad, leave it well alone. Toad equals toxic. They have lumps behind their heads called paratoid glands which are filled with nasty chemicals. The skin secretes these and – although it *is* possible to prepare some toads in such a way that they're ready for the pot – if you get it wrong these chemicals can cause pain, inflammation, blindness and even death. Not worth the risk.

If you've got yourself a frog, examine it carefully before you even touch it. If it's very brightly coloured, leave it alone. Although most frogs *are* edible, there are some which are potently toxic to humans. Most of these (like the golden dart frog, see below) have brightly coloured skin.

The golden dart frog is found in the Colombian rainforest and is thought to be one of the most poisonous creatures in the world. Its skin contains a very rare, very lethal poison. A single frog contains enough of this poison to kill ten people, and there's no known antidote. Some indigenous people have used the poison to tip their hunting arrows to make them more deadly . . .

Hunting frogs

Your best bet is to hunt by the water's edge at night – if you're lucky, you'll hear them croaking and the noise should bring you straight to them. Frogs can be timid, so move quietly and carefully. If you have a bright light, shine this towards the water. It serves two purposes. First, the light will reflect in the frogs' eyes, allowing you to locate them. Second, frogs, like rabbits, will become stunned by the bright light, making it less likely that they'll jump away as you approach.

You can use a sharp spear – traditionally called a 'gig' – to pin the frog; the split spear on page 166 would be a good tool for the job. Alternatively, you can use a net. Or, if you're fast, your hands. I've caught plenty of big frogs this way – you just need to be able to catch them faster than they can jump.

If it's definitely a frog that you've caught, and it's not brightly coloured, you can probably eat it – but you will have to skin it first, because even edible frogs have dodgy stuff in their skin that you don't want in your gut.

Before you skin it, though, you'll need to make sure your frog is dead. This can be harder than it looks, especially with large specimens. A knife through the back of the head should do it, though you might sometimes need to give the head a good hit with a blunt object to finish the job off quickly and painlessly.

Most of the meat on a frog is on the legs – you can cut these off and skewer them on a sharp stick. Hold the legs close to your fire and the skin will burn off, then you can cook the meat through. This works better for smaller frogs.

Alternatively, you can skin the whole thing. If you make an incision down its back, you should be able to peel the skin away from the body – a bit like pulling off his trousers. You should also find that this takes the guts away from the body too. What's left is fine to eat. Or you can cut off the head and pull the skin away, before gutting the peeled frog separately.

Frogs can be fried, grilled or boiled. You'll probably find yourself concentrating on the legs where most of the meat is, but on bigger frogs you can get a few decent bites out of the breast as well. In an emergency you can eat them raw – I've done this myself several times. But they're a bit more appetizing when they've been cooked, and it's always safer to cook your meat first in any case.

Frog Soup

This is a lot nicer than it sounds! It's good to make if you've caught yourself plenty of small frogs. Skin and gut them, then rinse them in fresh water. Bring some water to the boil and add a stock cube if you have one in your pack – this will add much-needed flavour to the soup, because the frogs themselves don't carry much. Simmer the frogs for about 10 minutes, then add any wild greens you have to hand and cook for another 5 minutes. The meat should fall nicely off the frogs, and the broth will be enriched with their nutrients.

SALAMANDERS

In China, the Chinese giant salamander – which is the world's largest amphibian – is considered so delicious (and is so easy to hunt, because of its size) that it's becoming endangered and it's now illegal to hunt it. But not all salamanders are edible. As with frogs, there are some which secrete a toxin from their skin, so be careful about touching them unless you're sure. Generally speaking, nature gives you a warning about these poisonous ones – they're very brightly coloured, which in the animal world is often a clear sign that you should stay away.

You can hunt edible salamanders in much the same way as you hunt frogs. In fact, it's a bit easier, because salamanders don't move so quickly. You'll find them by, or in, the water, but also hiding in warm, dark, damp places. Maybe you've heard the myth that salamanders are created in fire. They're not, of course, but the story probably comes from the fact that they sometimes hide out in rotting log piles – when you set fire to the wood, out they come.

Treat salamanders in the same way as frogs – make sure you remove that skin! The best way to do this is to skewer your salamander to a tree, or some other fixed point, and peel the skin off whole, using pliers and a knife. It's quite a tricky process and works best for larger specimens. If you've no pliers, just use your knife and your hands and do the best you can.

REPTILES

By reptiles, we mean, broadly speaking, snakes, crocodilians (that's crocs, alligators and caimans), turtles and lizards. It might sound like this class of animals doesn't offer much in the way of food. Think again. Reptiles have plenty to give us in a survival situation.

Trouble is, they're not always the easiest – or the safest – to get your hands on . . .

SNAKES

All snakes are edible, even the very venomous ones.

You hear people say that snake meat tastes like chicken, but I don't know what sort of chicken they are eating! Snake is much tougher, much bonier. Having said that, some snakes don't really taste too bad, and they're full of protein. I've eaten puff adders before that contained enough venom to kill five people, both raw and cooked. I can just about stomach it raw, but for something so aggressive and dangerous, the cooked meat actually tasted pretty fine.

Since reptiles are cold-blooded, they don't carry as many blood parasites as warm-blooded animals. This makes them a bit safer to eat raw. They can still contain some parasites, though, not to mention salmonella – so it's better to cook them if you can.

Snakes are a good survival food because you find them in a wide range of terrains, from grassland, to tropical jungle, to desert. They're found on every continent except Antarctica – although some larger islands, like Ireland and New Zealand, claim to be free of snakes.

But, of course, you have to know how to handle them – or *not* handle them, depending on the situation. And you must take the proper precautions before you try to tackle one. So before we look at how to prepare and cook a snake, here are a few pointers on how to catch them.

Catching snakes

The first thing to say is this: if you know a snake is venomous, **do not** mess with it unless you're very confident about what you're doing. It's a good idea to familiarize yourself with the venomous snakes of whatever terrain you're entering. This will give you a good idea of what to avoid and, conversely, which snakes to go after for food.

There's no dead certain way of telling if a snake is venomous, other than being able to identify it. But a very rough rule of thumb is that triangular heads tend to indicate venomous snakes. Constrictors and other non-venomous snakes tend to have heads that run in line with their bodies, with no triangular shape. But please remember, this is only a rule of thumb – the coral snake, for example, does not have a triangular head and is extremely venomous.

Generally speaking, the constrictors (that's boas and pythons) are easier to catch than non-constrictors. They're much slower and are not poisonous. However, you should avoid them if they are very large. They can be aggressive and give you very nasty, if non-venomous, bites. Not to mention that if a very large constrictor gets its coils around you, and you aren't that big a human, then it is able to squeeze the life out of you bit by bit, breaking every bone in your body before swallowing you whole. So choose your battles carefully.

I've been bitten by a green tree racer in the Borneo jungle before, and it isn't fun. I grabbed it by its tail as it raced along a tree branch and it flicked back and bit me. I pulled it off my hand and it then started biting itself, it was so angry. It all ended OK, though – I ate it for supper.

It's worth remembering that pretty much all snakes are scared of humans. They are very sensitive to vibrations in the ground and will hear you walking from (literally) a mile off. They'll hide under rocks, in holes, in the long grass, under wood piles – chances are you'll have to go hunting for them. A long pole with a hook at the end is a good tool for this – you certainly don't want to go sticking your hand into a snake hole. Use this to hook and drag them out into the open.

Remember that when it comes to handling snakes, if you control the head, then you control the snake. Obviously, it is only ever going to bite you with its mouth and this is where the venom sac always is. Pressing down at the back of the neck with a forked stick is the best way to control a snake, as this keeps it pinned in place and stops it moving its potentially most dangerous bit. You should do this even with non-venomous snakes, because they can still bite and cause nasty infections (there's an old saying: snakes don't brush their teeth!). Infections from a snakebite can do you a lot of damage in the field, especially in the jungle, where the warmth and humidity can cause bacteria to multiply extra quickly.

Once your snake is pinned down, the best way to kill it is to club it to death with another stick – use a heavy, swift, chopping action to the back of the head – or with a sturdy rock. You need to do your very best to make the first blow fatal – not only because it's more humane, but also because wounded snakes are aggressive and dangerous. Alternatively, swiftly cut its head off behind where you have it pinned.

Don't let your guard down just because you think you've killed it. Many snakes are good at pretending to be dead, so if you free them too soon you could be in for a nasty surprise. Even after they are dead, the body can twitch and writhe for some time and the head of a poisonous snake will still contain venom. It is even able to bite you after it is dead! That's why I always cut the snake's head off before handling the body, even if I have clubbed it to death beforehand. Do this with your knife while it's still pinned down. Once the head is severed, you *still* need to be careful – there could still be active nerves in there that cause it to bite. Bury or burn it. *Never leave a fresh snakehead lying around.*

If you think you've been bitten by a venomous snake, forget any old wives' tales about someone sucking out the poison. That just risks putting the venom into the bloodstream of a second person. And don't cut the wound to try to drain it – that will just speed up the rate the venom is absorbed into your system. Instead, keep calm, wash the bite and keep the limb low (below your heart level). Try to identify the snake and then get to a hospital, fast. As soon as that venom hits your bloodstream, you're walking around with a potentially lethal time bomb inside you. In the meantime, try to keep the part of the body that has been bitten as still as possible to stop the venom moving round your body. Remove any watches or jewellery from the limb that has been bitten because they can cut into your flesh.

Above all, get medical help. Haemotoxic venom is going to start destroying your muscles and organs. Neurotoxic venom is going to destroy your nerves and brain. Both versions can be deadly. On the left is a picture of a friend of mine after he was bitten by a fer de lance snake in a Central American jungle. The snake injected him through the eye of his boot with a hit of haemotoxic venom. Be warned, be smart and always have an evacuation plan in place if you are going to be in snake territory for long.

How to skin and gut a snake

Skinning a snake is not always necessary – see page 236 – but if you want to, here's how. Turn the snake so it's lying belly side up. Now make an incision along the entire underside of the snake. Once this is done, pull the skin away slightly from the flesh at one end, then, holding the snake in one hand and the skin in the other, peel it away in one big piece. (You might occasionally find this a bit tough, in which case use the tip of your knife to ease away the skin from the flesh.)

Once you've peeled off the skin, get your fingers into the incision and pull out the guts, much as you would with a fish. They should come out easily, and often in one whole piece.

If you can get a snake's skin off intact, you can use it as a water carrier. To do this, rather than making a long slit along its belly, cut around the circumference of its body about 2.5cm above the anal vent. Now peel the skin off in one long tube. Clean it very well, both inside and out. Scrape the exterior (which was the interior) if you like, but be careful not to tear the skin – then tie a knot in one end before filling it up. (There may be scent glands near the anal vent – remove these before eating the snake.)

Snake guts make one of the best trap baits, and there are few baits that work better for catching fish. Because the guts are so tough, you can catch several fish with one piece of bait.

Some ways of cooking snakes

I'm going to show you a few methods of cooking a snake. The first is best for smaller, thin snakes that you've skinned and gutted. These can simply be coiled around a stick, tied in place and held over an open fire. (Wire works best to tie them with, as it doesn't burn through.) Alternatively, insert one end of the stick into the ground at an angle, so the snake is held above the heat.

A good rule of thumb for making sure you have the right heat is that if you can hold your hand in the heat for 5 seconds but no more, then it is good to cook with but won't burn the meat.

Bigger, fatter snakes can be cooked in hot sand – this is a great way of cooking a whole snake in the desert. You don't need to skin or even gut it first. Light a fire in the sand and, when it's burned down, brush away the embers. Now bury the snake in the hot sand and cover it again with the embers. The skin will protect the meat as it cooks through. After a while you'll be able to hear the snake sizzling beneath the embers. Carefully lift it out of your bush oven with a couple of sturdy sticks. It'll look black and charred, but you can now cut away the skin to reveal the nicely cooked flesh beneath.

You can also cut up a skinned, gutted snake into segments and boil it with whatever other edibles you have to hand. Or you can simply fry the meat – stir-fried snake is very popular in China, as expected!

I have worked with some indigenous people who carry live rattlesnakes in their packs but with the mouths sewn shut. That way, the flesh doesn't spoil in the heat and they always have fresh meat ready for whenever they want it – all they have to do is kill the snake.

LIZARDS

Lizards can be found almost anywhere, but they are most common in warm climates – mainly tropical and subtropical regions. And I'm not just talking about the little critters you see on holiday. Lizards can be massive – up to 3 metres long – and many have strong jaws and sharp teeth, which can give you a nasty bite. It's pretty special coming across large lizards like that in the wild.

All lizards are edible, but two are venomous. These are the gila monster – a slow-moving, snake-like creature which can grow up to 60cm long and is found in the south-western United States and the Mexican desert – and the beaded lizard, which is found in Mexico and South America. Both are haemotoxic and, although they rarely kill humans, you want to be careful with them because they can make you extremely ill if they bite you.

You've also probably heard of the Komodo dragon. If you come across one of those bad boys, leave it well alone – not only because they're endangered, but because they're powerful, fast and will readily attack a human. Their saliva contains eighty types of bacteria, which can cause fatal septicaemia in their prey. When the prey's dead, they tuck in.

It's possible to catch large lizards with your bare hands – I've caught a few monitor lizards in swamps this way. One method is to grab them by the tail and hold them up, which stops them from being able to wriggle around. You can then stun the creature by swinging its head against a tree or the ground. Alternatively, strike the lizard with a long, heavy stick to stun it, pin it down and then despatch it with a knife in the back of the head.

Smaller lizards can be caught using a lizard noose. This is a piece of light cordage (fishing twine or wire is good, but a long piece of dry grass will do just as well – or even a piece of dental floss) with a running bowline (see page 260) at one end to form a noose. Tie the other end to a twig. Holding the twig, gently ease the noose over the lizard's head and yank it upwards. The noose will tighten and the lizard will be dangling by your cordage. Noosing is effective because, although the lizard might see the noose coming, it doesn't perceive it as a threat – unlike your hand, from which it would run away immediately.

If you're proficient with a catapult (see page 167), that can also be a good way to kill a small lizard. Or, if you hit it with a stick with lots of branches, that can stun it enough to allow you to catch it. Just don't let a small lizard get anywhere near a tree – once it starts clambering up there, you've lost it.

Iguanas are a very popular food in Central America, where they are known as chicken of the trees! But raw lizard meat is very tough. I've eaten small ones whole, but there's a risk of parasites and salmonella. You want to get them cooked. Most of the good meat on a large lizard is on its tail. You can just remove the tail in its entirety, turn it upside down and split it lengthways along the underside. Lay it skin side down on an open fire and the skin will protect the meat as it cooks. While the tail is cooking, you can scoop out the entrails and locate the liver – it'll have a little green bile sac attached, which you need to remove before toasting the liver on a skewer over the fire. It tastes pretty good – and is full of nutrients.

Another way of cooking a lizard is to boil it. Cut into its belly and scoop out the entrails (preserving the liver, if it's big enough, as above), then place your lizard in some boiling water and cook it until the meat is tender but not too soft. You can now simply peel off the skin to get at the meat beneath. In a survival situation, you might also want to drink the broth – it'll be warming and contain much goodness.

Alternatively, you can roast a gutted lizard on a spit, just as you would any other small animal.

Lizard Shish Kebab

If you've caught yourself a small lizard – I'm talking just 7–10cm long – you can simply insert a sharpened stick into the backside and roast it over a fire for 5 or 10 minutes. Make sure it's properly cooked through before removing from the skewer and eating it whole.

Being cold-blooded, lizards are more sluggish in the morning or the evening when the temperature is cool. That's the best time to go for them.

CROCODILES AND ALLIGATORS

Crocs and alligators look similar but are part of different families (caimans are a type of alligator) and they have slight biological differences. Alligators are only found in North America and China, whereas you'll find crocs in many tropical and subtropical regions.

They're aquatic, and crocs can be divided
into saltwater and freshwater – although
salties can and do live in freshwater
rivers. Some people say that crocs are
more dangerous than gators, but that's too
much of a generalization: they can both
be terrifying and lethal.

Crocs are a massive danger to
humans. Salties are the most aggressive. If they
attack you and drag you into the water, your chances of surviving the
subsequent death roll are almost nil. They can snap their jaws shut
with a force of one tonne per square inch – it's thought that this makes
a croc's jaws stronger than a Tyrannosaurus Rex's – and Australian
salties are the continent's deadliest animal. But even freshwater crocs
and alligators will go for you. So you don't ever want to be between a
croc or gator and the water.

Be aware that these creatures can grow to massive sizes. When
they do, their skin is almost impossible to pierce, and they are *fast*.
A blow from their tail can be almost as bad as a snap of their jaws.
They're also cunning. They'll lie in wait for you and watch as you go
down to the water to drink. If you go to the same place for a second
time, they'll strike. They have amazing camouflage and can hide
underwater for long periods of time waiting for an unsuspecting meal
to come their way. This makes them undoubtedly the ultimate and
most fearsome stealth predator.

I remember once swimming across an alligator-infested river.
I chose to swim underwater because crocs and gators sometimes
mistake your bobbing head for something else – like a bird or turtle.
Swimming underwater stops that happening, but it's still a dangerous
game. If the croc is big enough and hungry enough, then you are
toast, however many clever tactics you try to employ. (I've heard of
the strategy of going for the palatal valve in the back of a croc's mouth
if it attacks you, which in turn lets water flood into its lungs. But good

luck to you if you ever manage that!) There aren't many things these creatures are scared of, and you're certainly not one of them.

The death roll is done even by small caimans. This is where the croc or gator grabs its prey then rolls over repeatedly and violently. If it is a small caiman, this rips chunks of flesh off the prey. If it is a big croc, the death roll will rip off entire limbs. If you are lucky (or unlucky) enough still to be in one piece after being dragged into the water, the death roll will drown you. Then the croc will take you down and wedge you under some log in the murky waters and wait for a few days to 'tenderize' you, before ripping you apart in bite-size chunks to swallow whole. (Crocs can't chew, remember.)

Have I made my point that hunting crocs and gators is not for the faint-hearted? Good.

However, having said all that, they are edible too, and our brains are bigger! And if there's nothing else to eat, some fresh gator meat could be a lifesaver.

In a survival situation, you should only ever consider going for the smaller crocs or gators/caimans – nothing much more than a metre long. They're still dangerous. They're still fast. And they are still very powerful and can kill. But against one of those, you might just stand a chance.

When I killed a gator in the Louisiana swamps once, my strategy was to creep up behind it stealthily, get close, then leap on top of it and pin it down with all my weight and strength. You want to pin it just behind its head, straddling its body. But beware if it tries to death-roll free! If it succeeds, you're in trouble.

A safer way is to make a noose from sturdy rope and carefully loop it over the top or bottom jaw of the croc's mouth. Attach a long stick to the noose if you want to try this. The croc will often allow you to do it, as it will be focusing on you and not the rope. It will no doubt be poised for attack, and will hiss a bit throughout, but you can persevere stealthily. Once the noose is over its upper or lower jaw, pull it tight. Give the rope a tug and as the beast death-rolls it will wrap the rope around its mouth and force it closed. (An alligator's jaws are incredibly powerful when clamping, but not when opening.)

Whichever technique you use, once you have it secured and pinned, the quickest and most humane way to kill the beast in this situation is by severing its spinal cord, which means a knife driven strongly down through the soft tissue behind the back of the head.

It's also possible to set a crocodile trap. Here's how.

The best tool for this job is a large, sturdy metal hook, about the size of your hand. If you don't have one of these, then a short, sturdy stick will do – it needs to be big enough that it will stick in the croc's gullet – about 10cm for a small croc or gator. Tie the hook or stick to a long piece of rope and attach the free end to a tree on the water's edge. Now you need to bait your hook. Any meat will do, and the smellier it is the better, because it will attract your prey from a distance. Animal innards that you'd normally throw away are a good choice. Wrap them round your hook or stick so that it's completely covered. This baited hook now needs to be suspended above the water, so you'll need to drape the rope over an overhanging branch. Let the bait dangle close to the water – if it's too high, the smaller crocs won't be able to reach it. The idea is that when the croc goes for the bait, the hook or stick gets lodged in its gullet, trapping the animal. It won't necessarily kill it, though, so you'll need to be extra careful as you haul your prey in and despatch it using one of the methods mentioned above.

Once dead, you can skin, gut and butcher a croc or alligator in much the same way as you skin large game. You'll find plenty of good meat on the carcass, especially on the tail, where the meat is very tender (hack it off just behind the rear legs), as it is around the jaw (which aficionados consider to be the best bit). The leg meat can be as tough as old boots, but a rack of gator ribs is a real feast. Croc and gator can be grilled, roasted or boiled in all the usual ways. It's great cut into chunks, skewered and cooked over an open fire, and you can also use it in any of the wild casserole recipes in this book.

There's a lot of meat on a gator. If you've got too much, you can dry it to make alligator jerky in the usual way – see page 194.

Try to remove as much fat from the meat before cooking it as you can, as it will make the meat taste very unpleasant. You can rub the yellow fat from a freshly killed alligator into your skin – it will act as a mosquito repellant.

Grilled Alligator Legs

Most people go for the tail, where most of the meat is. But in a survival situation you can't waste anything. The meat on the legs is tough, but perfectly edible. Hack the legs off whole. If you want, you can skin them. This will give the meat a better flavour. Grill the legs over the embers of a fire until they're cooked through before eating.

TURTLES

There are freshwater turtles and sea turtles. They're both edible, and both good sources of nutrients – low in fat and carbohydrate, very high in protein, not to mention an abundance of vitamins and minerals to keep you healthy in the field.

Sea turtles are found in waters world-wide, with the exception of the polar regions. They were traditionally used as food sources for sailors on long voyages when fresh food was at a premium, because they could be kept alive on board and killed when necessary. They are very good to eat. So good, in fact, that most types of sea turtles are endangered. Sadly, they don't help themselves by returning to the same beach where they were born to lay their eggs. Here, they can be easily caught and the eggs themselves foraged. But it's illegal everywhere to hunt sea-turtle eggs, and you'd only want to do this if it was a genuine question of life or death. When it *is* a question of life or death, sea turtles can tip the balance.

Freshwater turtles are also on the decline, but they are still widely hunted in many parts of the world – North and South America, Africa, southern Europe, Asia and Australasia – and there are many types that are not protected, which means you can eat them if absolutely necessary.

Don't be fooled by a turtle's lazy demeanour. In the Everglades, I once caught a snapping turtle – and they can be vicious. They have an aggressive, snapping beak that could easily take your finger off, and they've been known to kill each other by decapitation. The best way to make a turtle harmless is to roll it on to its back, or hold it up by the tail.

You can fish for freshwater turtles using a hook and line – they tend to be more abundant closer to the water's edge, so don't cast too far, or use night lines as on page 91. Alternatively, you can simply drag them from the water (mind those jaws). Or you can make a turtle trap. Here's how.

Make a ringed wall of sticks pushed vertically into the ground. It needs to be three times longer than the type of turtle you're hunting and only slightly wider. At one end, you'll need to construct a hinged door that opens inwards, as shown. Prop the door open with a stick and

place some bait inside the trap – any kind of meat or fish will do. As the turtle enters, it will knock the stick over and, as soon as it's fully inside the trap, the door will close down and the turtle won't be able to escape.

To despatch a turtle quickly and humanely, you can kill it by giving it a blow to the head and then removing the head with your knife. (Be warned – some turtles will continue to move even after you've cut their heads off.) If the head has retreated into its shell, you can use a hook to pull it back out. Alternatively, if you have no hook, you can sever its arteries by inserting a knife beneath the shell and driving it into the turtle's neck. (Smaller turtles can be despatched by dropping them straight into boiling water.)

Once you've decapitated the turtle, you can hang it upside down to collect the blood. It's considered a delicacy in some parts of the world, but it has real value to the survivor as a source of nutrients and liquid. Some indigenous people have been known to let the turtle blood clot and coagulate, before adding minced-up turtle flesh and skin, the heart, lung and liver and whatever seasonings they can put their hands on, then frying the resulting mixture for a nutritious meal. This is actually pretty tasty!

Once your turtle is dead, you can butcher it by slicing open the belly, scooping out the guts and then cutting away the meat (avoid eating the head and neck, as these can be poisonous). The meat can be boiled, fried and generally used as you would any other kind of meat, and the skin is very edible, albeit chewy.

Turtle soup is a speciality in many parts of the world. You can improvise your own survival version by boiling turtle meat in a little water with a stock cube, removing any meat from the bone and returning it to the soup, then adding a few diced vegetables and boiling again till they're cooked. The resulting soup will go a long way to restoring both your morale and any vitamins and minerals you might have lost in the wild.

Alternatively, I have once placed a whole turtle, shell and all, upside down on to the embers of my fire. It took about an hour to cook. Do this, and you can tell when it's ready because the shell becomes brittle and will crack easily. Hack away the charred shell to get at the nutritious turtle meat beneath.

> If you've eaten a turtle or tortoise and the shell is still intact, don't throw it away. Use it as a cooking pot. Boil it first to make sure it's perfectly clean. It can now be placed directly in the embers of your fire, or you can use the hot-rock cooking method described on page 105.

> All tortoises – which you can think of as land-based turtles – are also edible, and they mostly taste really great. So great, in fact, that the Giant Tortoise of the Galapagos Islands is almost extinct because it was hunted so persistently during the last century. If you ever find yourself there, look for something else first!

REPTILE EGGS

Reptiles lay eggs. They're edible, but you need to be super-careful about collecting them because all animals will protect their eggs aggressively. So, while alligator or rattlesnake eggs will make you a decent meal, think twice about messing with them. However, if you do manage to forage reptile eggs, make sure you cook them well before eating to avoid salmonella or parasite infections. This can be done by boiling, or by baking them on a fire (see pages 26–7).

All eggs, whether bird or reptile, are great sources of protein. If the shell is uncracked they can remain edible for quite some time. If you plan to keep them, though, don't wash them clean. Bacteria cannot penetrate a dry egg, but if there is any moisture it can allow pathogens to pass through the shell. Also, many creatures apply a protective liquid coating around the egg, called a cuticle, which protects its quality. If you wash the eggs, you wash off the cuticle.

AFTERWORD

B eing able to find food is one of the cornerstones of all survival. Food gives us precious energy and – almost as importantly – boosts our morale. I hope that this book has given you some of the skills and all of the confidence that you need to get out there into the wild and start thinking about food in a different way.

You might have found some of the contents in this book truly extreme. (I know I wouldn't want to be eating the partially digested contents of a dead animal's stomach every day of the week.) But to finish off, I want to go back to something I said at the beginning. This book contains food-gathering techniques that you may only ever use when you find yourself in a genuine survival situation. (I certainly hope that you'd never be tempted to kill a single living creature in the wild unless you were intending to eat it or your life depended on it.) But you don't need to be trapping wild pigs or pit-roasting puff adders in order to take advantage of the amazing bounty with which nature provides us.

Even if you just boil up a few nettles after your next walk in the woods, or gather a few mussels when you find yourself by the sea, you'll be far more in touch with nature than most people ever manage to be.

The survival skills I try to teach people can take a lifetime to learn. Even now, I'm still learning myself, every day. But by gathering a little wild food and starting to understand how we can rely on nature (and not just supermarkets) to supply our food, you'll be taking the first steps on a path that I hope will teach you a great deal about the world around you.

In short, knowing how to gather food in the wild will definitely improve your life. One day, it might even save it too.

Bear.

APPENDICES

APPENDIX A

CORDAGE

Throughout this book you'll have come across instances when you'll need some sort of cordage in the wild. Fishing lines, making a snare noose, tying the bottom of a cleaned stomach to make an improvised water bottle, lashing poles together to make a drying frame, even making a bowstring – all these jobs need string, rope or twine. But there's a limit to the amount of this stuff you can carry with you, and in a survival situation you might find you have none at all. So here's how to make your own.

Anything fibrous can be turned into cordage. You can then plait this – if it's long enough – into sturdier lengths of rope. But you need to keep a few things in mind before you choose your material.

First, is it long enough? When you turn fibre into cordage, you end up shortening it. Second, is it strong? Give it a tug. If it doesn't snap, tie a simple knot in it and pull it tight. If it still doesn't break, your material should be good and strong. It also needs to be flexible, and not too smooth – the fibres need to grip on to each other.

The natural world can supply plenty of material that matches these criteria. Plant stalks (nettles are particularly good and common). Seaweed (if you haven't eaten it first!). Animal hair. The best cordage is made from the interior of dead tree bark – you simply peel it off in long strips, before separating the strips until you have multiple lengths of fibre of the thickness you need. The best fibrous material comes from willow and lime trees, but really there aren't many trees whose inner bark won't supply you with decent raw material for cordage.

It's worth soaking your fibre before you turn it into cordage. This stops it shrinking as much when it dries out as it would if it dried from its natural state. This process is called 'retting' and, once you've done it, you're ready to make your cordage. I'm going to show you two ways to do it: laid cordage and plaited cordage.

LAID CORDAGE

Laid cordage is made by twisting fibres together.

To do this, take a long piece of fibre and twist it in one direction until it wants to make a kink. Now fold in a third of the fibre. Pinch the fold, then put the fibre on your lap and roll it away from you with the palm of your free hand.

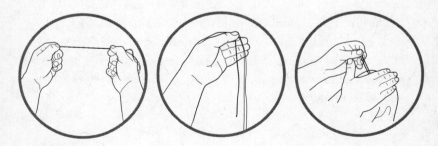

Keeping your palm fully held down on the fibre, release the pinched join. The cord will naturally twist from one end.

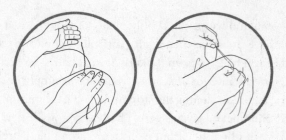

Now, pinch where the twist ends and repeat the process until you are about 5cm from the loose end. Lay another piece of fibre up to the loose end and continue the process – the new piece of fibre will become part of the old one.

You should end up with a good, strong piece of cordage, but you can make it even stronger by repeating the process.

PLAITED CORDAGE

If your fibre is not quite flexible enough to make laid cordage, plaited cordage is the way to go. It's just like plaiting hair – as the father of three boys, this is not something I get to do very often, I must admit!

Take three long fibres and tie them together at one end. Fold the left-hand strand over the middle one. Now fold the right-hand strand over what is now the middle strand (but *was* the left-hand strand). Repeat this process until you've plaited the entire fibre. Make sure you keep the plaits tight as you go. Tie or clamp the end when you've finished and let the strands dry out a bit before using. And remember that you can plait several of these pieces of cordage together for an even thicker rope.

APPENDIX B

KNOTS

In Appendix A, you learned about natural cordage. But to put any kind of cordage, natural or otherwise, to good use, you'll need a few knots up your sleeve.

There are loads of different knots you could – and should – learn. It's a hugely satisfying skill to have, and a potentially lifesaving one too, if you're going to spend any time in the wild.

When it comes to the business of wild-food survival, the truth is that you can do almost all tasks with just a handful of basic knots. On the following three pages, you will find five excellent, trusty knots – the best ones to cover a multitude of situations.

CLOVE HITCH

You'd use a clove hitch to attach a rope to a horizontal pole or post. For example, you could use it to secure the night line on page 91 to a fixed point on the river bank. A clove hitch is fast and easy to tie and untie, although it can come undone if it's tied to an object that rotates, or if there isn't a constant pressure on the line, or if the line is nylon and slippery. Add extra hitches to make sure the knot is secure.

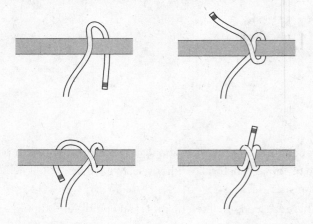

CONSTRICTOR KNOT

This is a great knot for tying the neck of a bag or sack. You might use it when turning an animal stomach into a water carrier (see page 185), or for tying the end of an intestine sausage (see page 186). It's similar to a clove hitch, but is hard to untie.

SLEDGE KNOT

This is an awesome knot for construction – but be aware it's so strong that once it is tied under pressure you won't be able to untie it. Ideal for tying a large animal's limbs to a pole, for example (see pages 182–3).

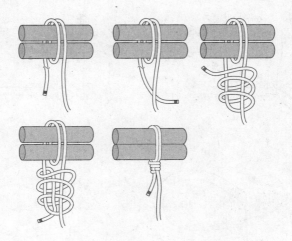

REEF KNOT

This is used for tying together two pieces of cordage of the same thickness – essential if you need to make a long piece of string out of several shorter pieces.

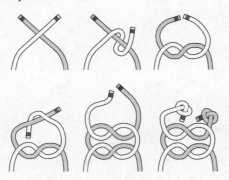

RUNNING BOWLINE

A good knot for snares – it will act as a noose and tighten when something pulls on the loop.

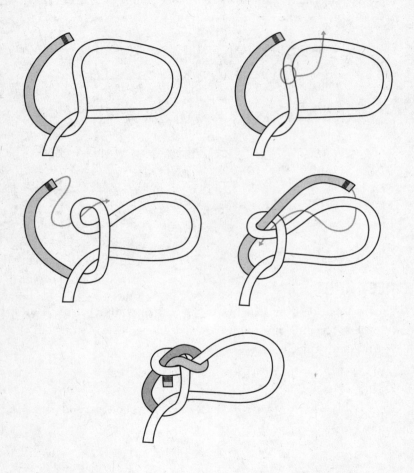

PICTURE CREDITS

The following photographs/line drawings were supplied by BGV: 161; HL Studios: 21, 23 (top and bottom), 24, 26, 27, 32, 63, 86, 87, 90, 91, 92, 93, 94, 95, 97, 102, 103, 105, 115, 134, 135, 136, 137, 138, 139, 147, 148, 149, 150, 151, 152, 154, 155, 156, 163, 164, 165, 179, 183, 244; Micaela Alcaino/TW: destroying angel (71), fool's funnel (74), beefsteak fungus (76), giant puffball (77), chicken of the woods (78), cauliflower fungus (79), sea lettuce (111), dulse (112), gutweed (112), carragheen (113), kelp (113), laver (114), running bowline (260); Patrick Mulrey: 255, 256, 258, 259; Shutterstock: fireweed (51), oleander (56), catfish (106), razor clam (119), limpet (120), sea urchin (121), crab (122), ant (204), fly (213), wasp (219), cicada (220), snake (232), sea turtle (243); Steve Rankin: 235.

Colour section
All images courtesy of Discovery except page 4 (bottom) and page 8, courtesy Simon Reay.

Every effort has been made to trace the copyright holders of images reproduced in this book. Copyright holders not credited are invited to get in touch with the publishers.

INDEX